Statistical Analyses for Criminal Justice and Criminology

This book is a how-to guide on statistical analyses designed for undergraduates and others new to the subject. It uses a conceptual framework, starting with the most basic concepts of statistics and moving up through the capacity to perform bivariate regression.

Written in an easy-going and clear style, it uses policing data to illustrate concepts. Easily identified Main Take-Aways and Key Terms features aid student understanding. Designed to combat the fear of mathematics and statistics often held by students in the social sciences, plain verbiage, multiple examples, and clear demonstrations combine to achieve the actualization and proper contextualized use of univariate and bivariate statistics. This work also serves as a launching pad for further study in statistics.

As an accessible introduction to statistics in criminal justice and criminology, this text will appeal to both students and instructors in introductory criminal justice and criminology statistics courses.

Michael J. DeValve is an associate professor in the Department of Criminal Justice at Bridgewater State University. His scholarly focus is on the nexus of justice and love. In addition, DeValve has written about justice organization capacity building and conflict resolution. He is the author of numerous works, including *A Different Justice: Love and the Future of Criminal Justice Practice in America* and *Personal Ethics and Ordinary Heroes: The Social Context of Morality* (Routledge, 2021). He has been teaching analyzing criminal justice data for many years using the concepts in this text.

Statistical Analyses for Criminal Justice and Criminology

A Conceptual Introduction

Michael J. DeValve

Routledge
Taylor & Francis Group
NEW YORK AND LONDON

Designed cover image: John L. DeValve

First published 2024
by Routledge
605 Third Avenue, New York, NY 10158

and by Routledge
4 Park Square, Milton Park, Abingdon, Oxon, OX14 4RN

Routledge is an imprint of the Taylor & Francis Group, an informa business

Library of Congress Cataloging-in-Publication Data
A catalog record for this title has been requested

ISBN: 9781032638584 (hbk)
ISBN: 9781032638546 (pbk)
ISBN: 9781032638591 (ebk)

DOI: 10.4324/9781032638591

Typeset in Sabon
by codeMantra

Access the Support Material: www.routledge.com/9781032638584

Table of Contents

Acknowledgments *vii*

1 Introduction and Basic Concepts 1

2 Presenting Data 10

3 Centrality 27

4 Variation and Dispersion 39

5 Basic Probability and the Normal Curve 54

6 The Central Limit Theorem and Sampling 66

7 Point Estimates and Confidence Intervals 77

8 The Reasoning Behind Hypothesis Testing 86

9 Bivariate Hypothesis Testing: *t* Tests 102

10 Bivariate Hypothesis Testing: ANOVA 116

11 Bivariate Hypothesis Testing: Measures of Correlation 127

12 An Introduction to Bivariate Linear Regression 146

13 Nonparametric Tests of Relation and Measures of Association 157

References 177
Appendix A: Proportions of the Area under the Curve, Normal (z) Distribution (measured from the mean, 0) 179
Appendix B: Critical Values of t and Degrees of Freedom 180
Appendix C: Area under the Curve, F Distribution, p = 0.05 values 181
Appendix D: Chi-Square Distribution Values 182
Index 183

Acknowledgments

At the risk of sounding like the walking Hallmark Card that I am, it is altogether meet to invoke the oft-used cliché that this is a work of one writer but many authors. My folks – Jeanne (who copyedited the manuscript) and John DeValve (who contributed the cover art), my beloved Lindsay, my wonderful child, my esteemed BSU and FSU colleagues and my deeply beloved and honored students, and also my dear friends Dave Wright and Xiaochen Hu played a role; to them I shall always be grateful. But the people to whom I simply must give special and humble thanks here are my math and statistics teachers. I have been extraordinarily fortunate in this regard. I don't remember the name of my undergrad stats teacher at SUNY Brockport College, but he was truly masterful. The three folks to whom I owe the most, though, are my statistics professors in graduate school, Steven Cuvelier and Charles Friel, and my high school math teacher, Don Van Derwerker.

I *hated* math class. Mr. Van Derwerker didn't varnish things or make math taste better. He taught with a quiet energy and resounding patience. He saw ability where I only saw drudgery, and he stuck by me; in the end, he did what few teachers of any kind could do: he taught successfully a thing to someone who was indifferent to learning that thing. Were it not for his rudderlike teaching and unswerving support, I would not have been remotely prepared to handle what followed. I hated math, yes, but despite my best efforts, because of Mr. Van Derwerker, I understood math.

Steve Cuvelier is one of *those* statisticians. He was decades ahead of everyone else, and I imagine he is still today. I learned more from a hallway conversation with him than I did from reading any chapter in any advanced stats treatment.

Chuck Friel taught me many things, but the lesson that stands out now is the one about elegance, offered by teaching elegantly. Overhead transparencies drenched in obscene detail in an era of PowerPoint slides, coupled with his commitment to unlimited availability, he guided generations of criminologists to statistical wisdom, and away from statistical floridness.

Thank you can't come close, sirs.

It has been decades since I have been a student in their classes, and I still hear each of these teachers in my head. I take it to be a categorical imperative for me to introduce you to these three master teachers to whom I owe so much. If you doubt your ability, if you hate math, if you think there's simply no way you can wrangle even a C out of a stats class, then because of teachers like them, think again. They're here with us, even if only in spirit, and they'll get you through if you only commit to the task. If you have no doubts about your ability, then because of teachers like them, we can go to work and go beyond.

Chapter 1

Introduction and Basic Concepts

Introduction

The criminal justice workplace is changing rapidly. The kinds of things you might be doing some years from now, the skills you will need, cannot fully be stated with any certainty. If you will work in some line-level job, problem-solving and conflict-resolution skills are almost guaranteed to be in demand. Almost as certainly, data analysis skills like those we will discuss here mastered to the point of flexible use across contexts and content likely also will be impactful. Whether we like it or not, the next decades will be influenced by "big data." The ability to handle data, to create data, to analyze and interpret results will be an increasingly useful and marketable capacity. Close attention must be paid as well to the considerable deleterious consequences of data-driven meaning-making (e.g., Buscariolli, 2023); people who work in policing data must be attuned to how they understand the manifold environments in which they operate. My purpose here is to present an introduction to those skills.

It is *only* an introduction, though, that I offer here. You will need to be ready to move beyond what is presented and deepen your skills and knowledge – your wisdom. My aim is to equip you with a sound starting point for launching your future mastery of analytical skills for criminal justice. I also want to give you some of my excitement for these and other analytical techniques. OK, yes, I know we are talking about statistics, and that most people don't find statistics "exciting," even a little. To be completely honest and fair with you, not even I, the author of this book on statistics, finds statistics to be *fantastically* exciting. I recognize the value of reasoned, sound analytical techniques, though. And as our society becomes increasingly reliant upon data-driven processes (like COMPSTAT and micro marketing), such analytical techniques will only increase in value. Just as surely, though, the techniques themselves must be employed wisely and ethically, not in a routine, habituated or disconnected way. The ability not only to use well-known analytical techniques, but also to adapt them to new settings and even invent new techniques altogether is a necessary skill for students of criminal justice and criminology as we move into the middle of the 21st century.

Before we really get into the work, I want to take a moment to tell you about the data that we will be using throughout this text. One of the datasets is from my own research, gathered with my colleague Dr. Elizabeth Wright. The Fayetteville (NC) Police Department (FPD) asked us to administer a survey to citizens in order to learn more about their satisfaction with service, fear of crime and some other things. The other data came from the Police Data

DOI: 10.4324/9781032638591-1

Initiative (PDI), available online to everyone at www.policedatainitiative.org. The PDI is a law enforcement community of practice that focuses on leading law enforcement agencies, technologists and researchers committed to engaging their communities in a partnership to improve public safety. The PDI originated as a result of several recommendations in the President's Task Force on 21st Century Policing and represents the great work and leadership of more than 120 law enforcement agencies. The PDI is affiliated with the National Policing Institute. Established in 1970, the National Policing Institute (NPI, formerly the National Police Foundation) is the oldest nationally known 501(c)(3) nonprofit, nonpartisan, independent research organization dedicated to improving policing in the United States. NPI supports changemakers in policing, communities and government by harnessing the power of science and innovation to promote public safety for all. The PDI is one of many projects led by NPI. Absolutely check them out at www.policinginstitute.org. This book would not be possible without their support, and I want to express publicly my gratitude and my appreciation for people and important ideas of both the PDI and the NPI.

The Analysis Decision Process

To be able to answer questions in our discipline, it serves us well to think like an investigator. Often investigators are taught to ask about what, who, where, when, why, and how: what happened, who is responsible, when did it happen, why did it happen, and how did it unfold and how do we know? Well, researchers operate in much the same way. We will need to think about a thing I'm calling the **analysis decision process.** It is complicated and not always does it flow in a straight line, but it is our first crucial tool for getting useful answers to real questions.

The process of conducting a human scientific research project, regardless of what type of analyses we plan to perform, involves the asking of a series of questions in advance in order to assure through-put and robust connection between the thing you or someone wants to know and what you actually end up investigating (Figure 1.1).

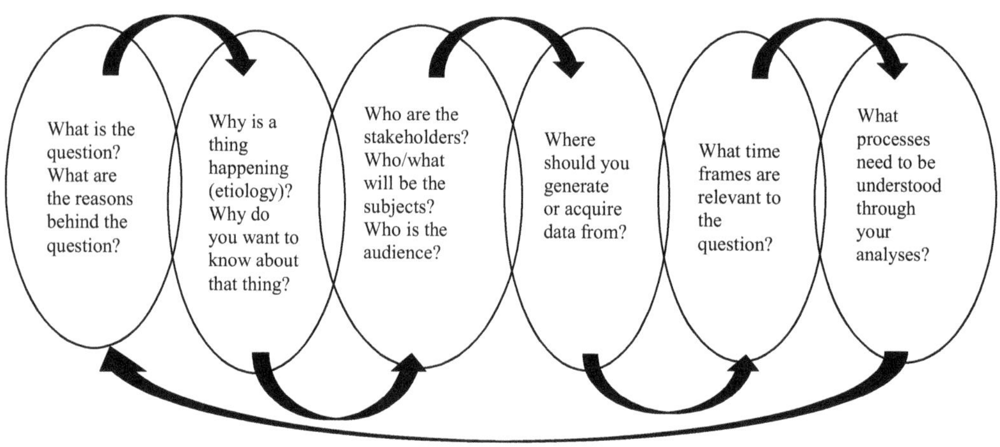

Figure 1.1 Analysis Decision Process.

What?

Ideas matter. Ideas shape how we understand the world, and indeed that is the whole point of all of this. Sometimes, the ideas we think are behind what we do are not always precisely the ideas that actually drive us or others. It helps us think about something called the "theory of action," then. Argyris and Schon (1974) enlarged on the "what-we-say-is-not-the-same-as-what-we-do-ness" of people in organizations. Espoused theories are those things we *say* are driving our action. Theories-in-use are ideas that *actually* do the heavy lifting. When there is a gap between espoused theories and theories-in-use, well, we have something of a challenge (see, e.g., DeValve, 2020a; DiMaggio & Powell, 1983; Meyer & Rowan, 1977). If you detect a gap between espoused theories and theories-in-use, care is certainly warranted. Sometimes it is a time for helping to educate stakeholders and sometimes it is better to be quiet and observe.

Here's the thing, though: ideas come together into theories. Some theories are grander and more formal, like those you might come across in a criminological theory course. Others are less formal, or even unstated altogether. Programs or policies, say, like park-and-walk practices for officers, have something we call **program theory**. Essentially, program theory is the idea behind the program or policy. Ideas live behind much of what we do, like the diminutive old fellow pulling the levers behind the Mighty Oz in *The Wizard of Oz*. Identifying program theory can be absolutely essential in the analysis decision process. And we forget to our peril that although the little dude behind Mighty Oz seemed to be the weaker of the two, in point of fact, ideas are truly mighty. Sure, action is idea rendered on the world, but ideas are the things that power mighty action.

In a pure research setting, if you were the researcher, you would be left to decide for yourself what theories were most interesting and worth researching. Perhaps that time will come if you want it to, but it is not now. The chief wants to know something, and you need to discover what that something is *as well as why she wants it*. The trick, though, is that what the chief *really* wants to know (what theories are in use) may not be precisely what she asked you for (espoused theories). Basically, we must find the question she wants answered. Again, in a research setting, this combination of ideas involving a theory or theories and particular outcomes we call a **research question**, or if we felt like it, we might create a statement, something of a $y = f(x)$ statement, called a **hypothesis**. Creating formal research questions or hypotheses helps clear up the logic of the question and helps isolate the particular thing we want to know. Researchers often create hypotheses simply to help frame exactly what they or others want to understand so that the results can be more easily understood. After all, research is just a formal kind of a conversation.

Why?

In direct relation to the question of what one seeks to discover is the question of *why*. There are two connected but distinct questions of *why*: *why* do we think something of interest is happening, and *why* do we care to know about it? The first question, the one about why a thing is happening, we can call the **etiological why**: we want to know about the cause of something, and *etiology is the study of cause* (more on all of that later). We often have theories about why things happen, or don't happen, and human science can help us understand whether our ideas about why things happen are any good.

The second why, the question of why we want to know a thing, gets back to the realm of policy issues, which we might call an **applied why**. Perhaps the chief is concerned about the failure of proactive patrol to reduce crime. Perhaps the reason she might want to know why proactive patrol failed is that she is accountable to taxpayers. Perhaps the mayor or city council urged her into a policy she felt would fail, and now she has to report back to city leadership.

Who?

Speaking of city leadership, **stakeholders** often identify themselves to you by their telling you what they want, or by signing your paycheck. But not all stakeholders will (or can) show up for meetings. Stakeholders are those people who have a specific, concrete claim on the nature and quality of your output as a researcher and analyst. Stakeholders often include the chief and the rest of the executive staff, and also folks from the department who come to you with questions, city council members and anyone who has interest or who is impacted by the thing in question. Community members who want answers to police department decision-making often are also stakeholders. Stakeholders may help make your data, they may manage your data, and they may ask for insights from your data. And crucially, *stakeholders are people impacted by decisions made on the basis of the strength of your analyses.*

Where?

Once you have a sense of what answers you are chasing and for whom, the next thing is to figure out where you can find them. Often that means finding the right data. Now, let's be absolutely clear about something: when we talk about the "right" data or making data just the way you want them to be, we do not mean that you will be making data provide the outcomes you or your stakeholders want. Biasing results or falsifying data is the cardinal sin in research. *Bad* stuff. What I mean when I talk about "right" data or about making data according to our "preference" I mean that we are seeking to have *the most authentic and precise* answers humanly possible, and not data that serve some lesser agenda.

This is a key point: **the nature of the data will tell us how we should use them.** As data come ("data" is a plural word) in many forms, we need to become masterful at figuring out what kind of data we have and/or what kind we need. Sadly, sometimes what we have and what we need are wildly different things.

As researchers, we can create data (often called **primary data**), or we can use someone else's data (often called **secondary data**). **Agency data**, like that which we will often handle here, are something of a combination of the two, at least from the perspective of an agency researcher. It is like secondary data in that you yourself didn't get to determine exactly how the data would be made and maintained, so it may not be ideal for your purpose. At the same time, "yous guys" (or "y'all" if you're in the south), meaning *you the agency*, made the data, so it isn't totally secondary data. Maybe down the road, you will have a chance to increase data quality there or help those who actually record data do a better job at it for multiple purposes, but that's a conversation for another time.

Most often when we talk about data, we are talking about spreadsheets containing information. That information can be **quantitative** (or presented as numbers) or **qualitative** (presented in, well, non-quantitative ways). Numbers are very powerful symbols and as we will see soon, they can be used in different ways, but as we are talking about data most often contained in computers, we need to remember that, generally speaking, computers handle numbers far better than other stuff. That's mostly because they "think" in numbers. Thing is, sometimes qualitative data, like interview text or photos, or other kinds of non-numeric information can give us incredibly powerful insights, well beyond anything a dataset and sophisticated multivariate analysis could manage.

Today, most large justice agencies hold huge piles of quantitative data. To manage those piles, they are actually kept ... in piles: piles that are connected to each other. **Databases** are essentially connected piles of data: a "people" pile may connect to a "transaction" pile that may connect to an "arrest" pile ... Databases *should* be connected according to very precise rules, often called database theory, or database **normalization**. Normalization means that the organization of the data in the database complies with five basic conditions, known as **normal forms**. The purpose of normalization is to reduce repeated information while retaining as much information as possible, as accurately as possible, so that we can get the information we need when we need it, and *only* the information we need. We will not be analyzing databases themselves, so conversations about databases and how to use them are reserved for other treatments. What I do need to note is that it is likely some of the quantitative data we will analyze will come from a database in the form of a **queried report** in the form of a **spreadsheet**.

A spreadsheet has a very specific anatomy: it has rows and columns. Most of the time each row, or horizontal set of boxes, represents one object of interest, one record, one thing we are counting. We call that the **unit of analysis** of the data. If the spreadsheet was a report of officers for human resources purposes, each row would represent one and only one police officer – one *person*. If the report was about each patrol zone and its activities, then each row represents one and only one *zone*, where many officers might work.

Each column of a spreadsheet represents one field or **variable**. The things of interest, the units of analysis, have characteristics, and most likely those characteristics of interest vary: police officers have different color hair, different heights, different certifications, skills and such. So, say, "education" would be a variable for police officers. Officers differ regarding their educational attainment, and for personnel decisions, it helps to keep track of that. Well, as educational attainment varies, each possible answer to the question of "What is your highest level of educational attainment?" can change. Each discrete possible answer to the question is called an **attribute** or a **level**. Options provided for educational attainment might include "some college," "college degree," "some graduate school" and "graduate degree." Sex (male, female, non-binary) has three basic attributes or levels (excluding non-response; note also that a Gender variable may have many attributes or levels). Sex is a variable, and male, female and non-binary are the attributes or levels.

Those things, those attributes and their variability, are the very thing that makes statistics useful. Basically, we want some variability. *Statistically speaking, variety is more than the spice of life: it is life itself.* If all the officers in a patrol zone were female, data on officer activities within the zone would show sex not varying. Sex would be a **constant**. Constants like this are of almost no statistical use.

But not all variability is interesting or useful. In fact, take out your cell phone for a second, and look to see how many signal bars you have right now. These bars are not terribly precise, they differ by phone maker, and they may even consider network issues, too, but in a rough sense, they tell us how much signal and how much noise that phone is "hearing." Engineers talk about phones (and other communication devices) in terms of their **signal-to-noise**

ratios, or how much useful information a thing receives compared with how much static and other useless input it hears. We will talk about signal-to-noise ratios throughout the text, as much of what we will be doing can be seen in terms of signal and noise.

When?

Oftentimes, questions relate to when events occurred, how long they took and so on. The reasoning systems used in statistics are relatively straightforward affairs (meaning that they don't factor in subtle variations in meaning, as we do, say, when we look at other people's faces and interpret their feelings). These systems rely on a homogenization of objects (i.e., quantification) into numbers, and it just so happens that analytical software like Microsoft Excel, R, SPSS, SAS, Statistica, and the like handle quantitative data very well (because, of course, they were designed that way). These software programs can give us useful answers regarding patterns in quantified objects, or even predict when they might occur again. That is, as long as we have entered the data properly and told the software exactly how to use those data for our purposes, we can learn something (but almost certainly not everything) of value. We will talk about how to handle date variables later in the text, but before we do, we need to become sensitive to something truly vital.

A properly formatted date field in a database versus typing simply "Thursday" in a spreadsheet are far from operating the same way for our use. With a date, like a birthday, we can use a simple algorithm to find out how many days old someone is, but that kind of precision is not possible if all we recorded was that she was born on a cloudy afternoon, or that it was a Thursday in October. Measurement precision and consistency are crucial concepts, key to being able to operate with data.

Levels of Measure. In fact, this particular idea is so "key" that without mastery of it, useful statistical analyses are not possible. That idea is something we call **levels of measure**. Numbers are powerful symbols, only made more powerful because of how much we rely on computers. But because we use computers for our analyses, we often find it easier to use numbers to stand for things that are not naturally numeric. Sex, for example, viewed with three levels as we did above, is not a quantitative thing. But to make the variable easy for computers to handle, we assign numbers to each of the possible attributes of that variable; perhaps Female = 1, Male = 2, and Non-Binary = 3. We would call this kind of variable "nominal" because the number represents nothing more than a place-holder, a name of a kind.

Now the nature of each variable may or may not be measurable with great precision. Maybe we can only place the attributes in some kind of order (like, say, different levels of educational attainment). Maybe the attributes of our variable do not allow for any meaningful ordering at all (like with sex or, say, religious affiliation). It turns out the orderability and precision of a variable is one of the most important factors in determining what analytical techniques can be used on it. Variables whose attributes or values can only be placed in order, say, officer rank (officer, senior officer, sergeant, lieutenant, captain...) are called "ordinal." We can place them in an order, but the distance between officer and senior officer, perhaps coded 1 and 2 respectively, is not an equivalent distance between senior officer and sergeant, perhaps coded 2 and 3. The numbers only represent an order: a 2 is higher than a 1. If we were to talk about actual years of service, now that's a different matter altogether.

Let's say you had to determine who was to get department longevity awards at the annual PBA banquet. Would you simply assume that those with the highest rank had been around the longest? Doing so could get awkward; some people choose to stay at the officer rank their entire careers. Rank won't serve. You need to know how many years each officer has served. If HR isn't already keeping track of service length, you can secure each officer's hire date and do a little arithmetic. That variable, "years of service," has values that are orderable, it is true, but the value for each officer is *actually counting something standard*: years (or if you wanted to be really picky you might be able to calculate days) of service. We would say that this variable and similar variables have an "interval" level of measure, as the numbers it presents actually *do* represent meaningful intervals. We can subdivide the numbers (for example, someone can have exactly seven years of service, and someone else can have 7.5 years of service), make comparisons across groups, and use the numbers there in every way numbers can be used.

There's one more level of measure to know: "ratio." Now, because we are talking about years of service, no, not many are likely to have served for zero years, particularly if we are talking about service awards. Perhaps there are some brand-new officers who have yet to finish their first year, in which case, the number "0" *is* meaningful. When an interval variable has a zero value that is meaningful, we call it "ratio" because we can calculate ratios from it. To be sure, though, no serving officer will be 0 years old. Age, then, would be interval, but not ratio, at least in this example.

But it is worthy to note that interval and ratio-level variables do not differ grandly from each other *mathematically*. They only differ in terms of how we might understand zero values. If we are being perfectly pure, there *are* things one should not do with interval variables that one can do with ratio variables (like create ratios, or use statistical techniques that employ multiplication and division, but even this deserves a caveat). Again, though, for general purposes, there is little meaningful difference between interval and ratio variables. SPSS, for example, does not distinguish between them (it calls them both "scale"). This last point raises something that we will need to attend to briefly: the issue of varied nomenclature. Nominal, ordinal, interval and ratio are the four levels of measure, but as scientists seem to like having multiple names for things, they have created other names for them, and even grouped them for their use. Often one might hear the term "categorical" variables. The term refers to nominal and ordinal variables, usually variables that denote categories. The sister term is "continuous," which refers to (you guessed it) … interval and ratio variables. They are called continuous because they can be placed on a number line, meaning that the values there can be **subdivided** meaningfully, both along (i.e., in terms of greater or lesser values) and into (i.e., in terms of increasing precision) the number line.

If you're a little confused, that's just fine. Levels of measure is a vital concept and it might take a little time to get comfortable with it. Don't worry, though. We will be talking about levels of measure repeatedly all through this book. In fact, it turns out that the determination of the level of measure for any variable is actually quite simple. We can think of it in terms of a two-step, or two-question, decision tree. First, we ask, "Is the variable orderable?" If it isn't, it is a nominal variable. If it is, move on to the second question: "Is the variable subdividable?" If it isn't, it is an ordinal variable. If it is, it is a scale variable (Figure 1.2).

How?

It may be clear already that you are going to be asked to help leadership (like lieutenants, captains and the chief) make decisions. Often the decisions they need to make are difficult for a variety of reasons, not the least of which is that people's wellbeing is at stake. It makes

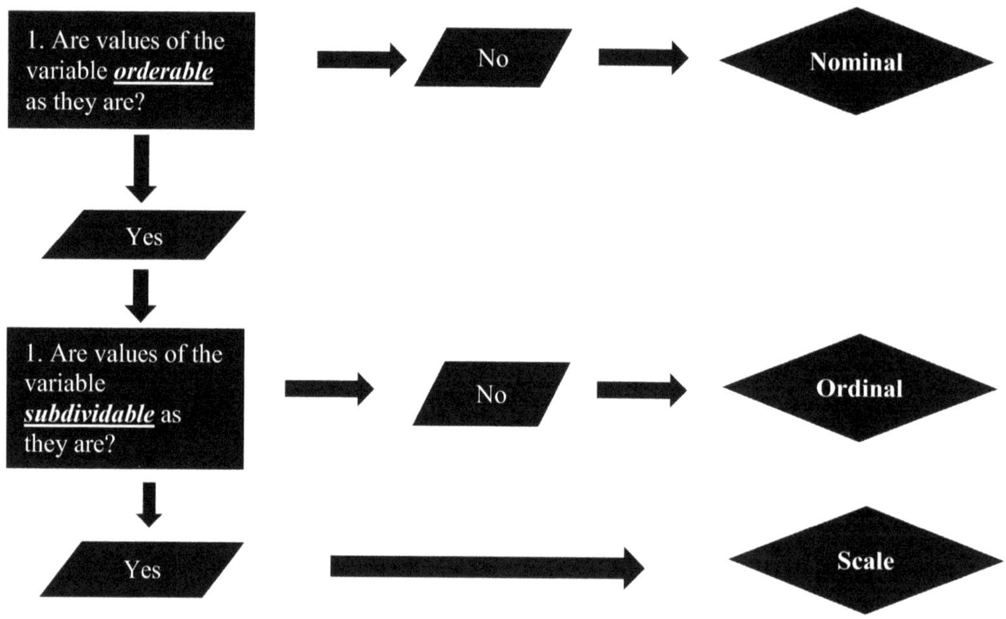

Figure 1.2 Level of Measure Decision Tree.

sense that they make those decisions using the best information available to them. That's where you come in. See, the data alone are not enough for making decisions. Some kind of **analysis** is necessary in order to leverage data for decision support. Often analysis involves the making of some kind of **comparison** (even in contexts other than statistical ones; see, e.g., Hallberg [2009]) in a **context** (or framework). **Analysis often involves the determination of the amount of signal relative to noise and to understanding that signal through comparisons made in some meaningful context.** You, as the analyst, choose the context and the shape of comparisons; you set the rules of engagement and thus, you can make results that are meaningful, misleading or downright nutty on the bases of the choices you make.

That means there will be times when you need to make some kind of decision, or some kind of referee-like call regarding the comparisons you make. It stands to reason, then, that there should be some guiding process or **criteria**, for making those calls. Referees, like the referee in a boxing match, utilize criteria, or decision-support frameworks, to allocate points or to determine if a bout should be stopped. The boxing analogy will turn out to be more apt than it might seem right now, but right now it is enough to say that you will need criteria for helping you make the referee calls you will have to make. These criteria are a crucial part of statistics and other analytical techniques you may use. This text will help provide just such tools for you.

All this being said, though, it is *absolutely essential* to remember that statistics and other analytical techniques themselves *do not make wisdom*. A guitar sitting in the corner can be used to make classical music, blues, free-form jazz, Dream Theater riffs or just noise. But the music comes *only from you or from me*. The technique to play the guitar is like the analytical tool you choose – it gets you the kind of outcome you seek, but *the music* is all from your head, your hands and your heart. There is no magic in analysis.

The magic – meaning the real life-improving insights – comes from what we are able to do with results. The magic, then, is you.

Conclusion and Review

Statistical reasoning is a powerful human tool for improving understanding of the human world. It is, like any tool, not to be mistaken for the thing made with the tool, though; a ship saw and a ship are two very different things. Human scientific practice has turned the corner; statistical results are not themselves insights. They are akin to a ship saw, perhaps, but ships do not come from ship saws alone; wisdom and policies and practices that come from wisdom are the ships we fashion.

The process of human science is not appreciably different from the process a police investigator might use to investigate a crime. Asking key questions, like what, why, who, where, when and how helps us be assured that the scientific investigations we undertake are meaningfully connected to the things we want or need to know.

Main Take-Aways

- Statistical reasoning is just that: a reasoning system. It is not itself a revelatory process for wisdom. The wisdom is yours.
- Because statistical reasoning is only a formalized problem-solving or question-answering process, it has requirements in terms of the quality and shape of the information we feed into it. Thinking of statistical tests as decision-support systems helps to illustrate the importance of the stuff we do both before and after we actually perform a statistical test.
- Information comes in many forms. Some information is readily usable for statistical analyses, but that doesn't mean they are what you need for *your* statistical analyses.
- The concept of levels of measurement (nominal, ordinal, interval and ratio) of variables is one of the most crucial ideas in this text. The sooner you can differentiate correctly among the four (or three, or two...) levels of measurement, the sooner you will be more adept at selecting the statistical test you need in order to answer your question.

Key Terms (in order of appearance)

analysis decision process
program theory
research question
hypothesis
"etiological why"
stakeholders
primary data
secondary data
agency data
qualitative
quantitative
databases
normalization
normal forms
queried report
spreadsheet
unit of analysis
variable
attribute/level
constant
signal-to-noise ratio
levels of measurement
nominal
ordinal
interval
ratio
scale
categorical
continuous
subdividable
comparison
context
criteria

Chapter 2

Presenting Data

One can be forgiven for thinking that "agency statistician" or "researcher" might be the most boring of offices one might have in a police department. Nothing could be further from the truth, of course. As a statistician, you are responsible for framing and shaping raw information for consumption, not only by rank and file, but by policymakers, their bosses, and *their* bosses, the public.

OK, so I can hear the next question bubbling up within some of you: "But I thought statistics were about objective truth – science and all." Well, objective truth is a rare thing not often found in the human sciences, and it is a thing rapidly losing some of its cache even in the physical sciences. Still, if you want certainty, Dr. Smith's physics class, and Dr. Jones' classics course are both just down the hall. *This* is not the place for certainty. And although we have lots of formulae and arithmetic, believe it or not, criminological statistics is not a wholly mathematical endeavor. For quantitative analyses, the math is necessary, but it is far from sufficient. Recall from Chapter 1 our point about statistics not being wisdom? Statistical analytical techniques can, when results are understood in local and wider contexts, *help us to* find *wisdom* by gesturing toward one set of ideas over another. But certainty is not a thing we seek here. Most often, certainty gets in the way of wisdom.

Basic Data Presentation Techniques

Mark Twain is popularly thought to have said, "There are lies, damned lies, and statistics." If indeed he did say that he was not wrong. Statistics are about analysis, about insight support, and ultimately are a form of shorthand communication. What, precisely, is said can be manipulated with ease to give wildly different understandings.

Even the most basic means of presenting data constitute a kind of analysis. That means that you need to have a wider sense of context when you begin to do even the most basic data presentation preparations. Let us turn our attention to some of the basic tools at our disposal for data presentation.

Often the first step in the process of creating most kinds of basic presentation of data is the grouping of data. Here (Table 2.1) we have some raw data, data that have not been manipulated in any way. These data were derived from the City of Detroit's Open Data Portal (2020). What we see here is merely a tiny and unsystematically chosen smattering of nine records from the more than 200,000 similar records in the city's database. Each row is merely a record of a particular event – in this case, each row represents a charge made for burglary. Columns, or fields, are variables to the researcher: "_charge_description_" and "_scout_car_area_" are two examples. We can see among the fields/variables what appears

DOI: 10.4324/9781032638591-2

Table 2.1 Screen Grab of RMS Crime Incidents: A Sample of Burglaries

crime_id	*charge_description*	*incident_timestamp*	*scout_car_area*	*precinct*
3001737	BURGLARY - FORCED ENTRY	12/9/2016, 7:00 PM	803	8
3004757	BURGLARY - FORCED ENTRY	12/21/2015, 4:00 AM	414	4
3094818	BURGLARY - FORCED ENTRY	7/24/2017, 5:47 AM	1103	11
3000566	BURGLARY - FORCED ENTRY	9/9/2016, 6:00 PM	906	9
3338066	BURGLARY - FORCED ENTRY	3/27/2019, 3:00 AM	401	4
3004732	BURGLARY - FORCED ENTRY	12/21/2016, 3:00 PM	1212	12
3096510	BURGLARY - FORCED ENTRY	7/27/2017, 9:52 PM	802	8
3035494	BURGLARY - FORCED ENTRY	3/8/2017, 12:11 AM	501	5
3073750	BURGLARY - FORCED ENTRY	6/5/2017, 1:00 AM	1102	11

to be the unique identifier, or key, record, called "_crime_id_" as well as the charge description, the timestamp for the incident, as well as some coded information about the police unit and patrol area involved. The unit of analysis for the data from which these cases come would be something like "criminal event," then, if we reason that the key field, recording unique things, seem to be counting crimes or crime events. A particular arrest may involve multiple charges, and tracking each of the charges brought becomes important for its own reasons. Looking at the rest of the fields, our suspicion is borne out; each variable (that is, each column) seems to record something relevant to a particular criminal "moment" in time. Details of the *what*, the *when*, the *where* and the *who* are recorded in columns for our use.

For all their detail, not much can be discerned from these data beyond the bare information related to each event. Something more is needed to derive some sense of clarity about what is happening here. Perhaps some form of organizing of cases is in order. In Table 2.2, we have a frequency table derived from raw data just like those records above. Instead of Detroit, Michigan data seen in Table 2.1, Table 2.2 shows property offenses known to DeKalb County Alabama Sheriff's Office for several years. The kind of information that helped make Table 2.2 would have looked much like what we see in Table 2.1.

Again, Table 2.2 was derived from data that looked much like the raw data shown in Table 2.1, but it has been organized in such a fashion as to present the number of events that correspond to each cell. Specifically, instead of a set of recorded values, like we might have in the raw data in Table 2.1, the frequency table (Table 2.2) presents counts of events that fall into particular categories indicated by row and column titles. In 1994, for example, we can see that there was a total of 250 property crimes in DeKalb County. Of these 250, 83 were burglaries, 160 were larcenies and seven were motor vehicle thefts. Instead of raw data, then, what we have in Table 2.2 is what we call a **frequency table**.

A frequency table is a rudimentary form of analysis because it seeks to impose an organization on raw information in order to learn something from it. It gives us some measure of context, some framework for comparison. We are pattern-seeking beings, and analyses, like the creation and interpretation of a frequency table, represents a means of providing context-bearing, pattern-showing organization. The data here in this frequency table are a somewhat more concise, or **reduced**, version of the information in the raw data spreadsheet, but how we organize and present that information, as well as how we read the resulting table, are all kinds of analysis.

Table 2.2 Property Crime Reported by DeKalb County Sheriff's Department, Alabama, 1985–2014

Year	*Population*[a]	*Prop Crime Total*	*Burglary*	*Larceny*	*MV Theft*
1985	53,922	331	162	131	38
1986	53,642	380	183	148	49
1987	53,927	326	168	122	36
1988	54,164	320	191	100	29
1989	54,460	189	93	82	14
1990	54,763	104	42	45	17
1991	55,342	172	99	64	9
1992	55,845	192	105	80	7
1993	56,484	301	183	114	4
1994	57,533	250	83	160	7
1995	57,285	406	159	241	6
1996	57,374	264	101	158	5
1998	58,274	0	0	0	0
2004	66,846	271	98	126	47
2006	67,829	674	358	210	106
2007	68,653	686	187	408	91
2008	69,291	535	168	310	57
2009	69,781	553	237	285	31
2010	71,159	0	0	0	0
2011	71,365	908	320	515	73
2012	70,944	937	390	471	76
2013	70,912	801	333	395	73
2014	70,993	810	312	427	71

Source: UCR (https://cde.ucr.cjis.gov/LATEST/webapp/#/pages/explorer/crime/crime-trend).

[a] Source: https://www.google.com/search?client=safari&rls=en&q=dekalb+county,+alabama+population+1985&ie=UTF-8&oe=UTF-8.

Actually, it is not too much of a stretch at all to think of a table like Table 2.2 as telling something of a story. In fact, this table contains a few stories. To begin, we can see that there appears to have been some change in the overall amount of property crimes, as well as which of the three types of property crime were most prevalent in any given year. The eagle-eyed among you might also have noticed that for two years (1998 and 2010), no property crimes were reported at all. You might also have noticed that not every year is represented in the data. I did not modify the records – what you see is precisely how they came (well, I took out a few columns and added the population column). Don't forget: UCR reporting is voluntary. And there is at least one more note to make at this point: UCR data did not include information about county population for each year. I had to go find it and pop it into the table (statisticians call that **imputing** data).

Again, the "reduced," or aggregated, information in Table 2.2 is a little handier than the raw information presented in Table 2.1 – we can do more with it. Specifically, there are some things we can do with frequencies that help us tell those stories in the data. In short, we can derive and provide some sense of context.

As I said, we can tell many stories from these data, and a key point for us to keep in mind is that not all of these stories are necessarily worth telling. Some are, and one of the purposes of this book is to help you determine which stories are worth telling and which are not. First, though, let's look at a few ways we might create and impart the stories we find.

And just one more point before we begin telling each other stories: as I said earlier, most commonly, the stories we tell with statistics involve some kind of **context**, and often, that context is achieved through a **comparison** of some kind. The stories that matter make meaningful comparisons and generate sound context for contemplation and action. The stories that deceive use misleading comparisons to offer false or imprecise contexts. **The smart consumer of statistics will keep an eye out for the comparisons made or invited, and the context generated by the comparison. The wise statistician will construct comparisons that highlight contexts that work to diminish human suffering.**

Cumulative Frequency

Sometimes, it serves our story to think about the total number of events that interest us as they accumulate over time. DeKalb County might be asked about the total number of property crimes over the time period in question for the sake of adjusting property insurance rates and property values. Cumulative frequency creates context by comparing each row's value to a total running tally, and to a grand total.

From Table 2.3, we can see that between 1986 and 2014, there were 9,390 reported property crimes known to the DeKalb County (Ala.) Sheriff's Office. This number is both precise and somewhat not. Again, let us remember that this cumulative frequency, or running total, is incomplete, as both data and years are missing from the available record. It is precise, though, in that it is the best official data we have, representing the most complete

Table 2.3 Property Crime Reported by DeKalb County (AL) Sheriff's Office and Cumulative Frequency

Year	*Population*	*Prop Crime Total*	*Cumulative Freq.*
1985	53,922	331	331
1986	53,642	380	711
1987	53,927	326	1,037
1988	54,164	320	1,357
1989	54,460	189	1,546
1990	54,763	104	1,650
1991	55,342	172	1,822
1992	55,845	192	2,014
1993	56,484	301	2,315
1994	57,533	250	2,565
1995	57,285	406	2,971
1996	57,374	264	3,235
1998	58,274	0	3,235
2004	66,846	271	3,506
2006	67,829	674	4,180
2007	68,653	686	4,866
2008	69,291	535	5,401
2009	69,781	553	5,954
2010	71,159	0	5,954
2011	71,365	908	6,862
2012	70,944	937	7,799
2013	70,912	801	8,600
2014	70,993	810	9,410

approximation property crimes of which DeKalb County Sheriff's Office is officially aware during this 29-year period.

Ratios and Proportions

Ratios and proportions are measurements that derive their context through the comparison of two terms. Typically, the comparison process is brought about through division. Ratios and proportions, similar though they may seem, are different things.

Ratios seek to create context by setting two values in relation to one another. Ratios compare one group to another group, one thing to another thing. We might talk about the ratio of police officers to citizens in a community, or the ratio of correctional staff to internees. The ratio of citizen complaints per year received by each officer might also interest us. We could think of ratios as *part-to-part* comparisons: correctional staff and internees are all humans though they perform different roles within the prison.

Let's say we wanted to contemplate complaints against officers received in our city of 10,000 shiny, happy residents during a given year. Let's also say our department has 31 sworn staff, seven of whom received complaints. Finally, let's say that our department received 121 complaints, 11 of which were sustained. To begin, we could identify the ratio of people to complaints: 10,000 becomes the numerator, and 121 gets slotted in as the denominator. This, of course, creates an improper fraction that will give us a whole number. You'll see that 10,000/121 = 82.64 citizens for every complaint. We could express the same relationship by flipping the terms: 121/10,000 = 0.0121. That tells us that each citizen lodged just a little more than one-tenth of a complaint during the year. This latter framing, although expressing the same relationship as the former, seems less clear, less sensible. Next, we might note that there are 31 sworn staff. If we placed the number of complaints (121) in the numerator and the number of officers (31) in the denominator, we would find that the department received 3.9 complaints per officer. Notably, though, not every officer received a complaint, so this ratio can be misleading. In fact, if we subtract the number of officers who received complaints ($n = 7$) from the total, we get 24 officers. If we divide 24 (the number of officers who did not receive complaints) by seven (the number of officers who did), we find that for every one officer who received a complaint, there were 3.43 officers who did not.

If we refine things a little further and consider only the officers who received complaints ($n = 7$), we get: 121/7 = 17.29 complaints. That means the department received 17.29 complaints for every officer who received any complaints in this year. We know well that not every one of these seven officers received the same number of complaints, and more, a considerable number of the complaints seem to have been unsubstantiated. More interesting, then, might be to talk about the number of substantiated complaints per officer who received at least one: 11/7 = 1.57 complaints. Now, all of this ratio business is separate from the detailed recognition that, say, two officers in our department are responsible for 6 of the 11 sustained complaints. Clearly, a more detailed, casewise treatment is *absolutely* necessary for training and personnel decisions as well as from an agency accountability perspective, but if we are to compare this year to other years, or this department in this year to other peer agencies elsewhere, these ratios can be of some value.

Proportions, on the other hand, are *part-to-whole* comparisons, comparing a group to a larger whole to which the group is member. We could want to know what proportion of community members play cards on a regular basis, or perhaps what proportion of college students major in criminal justice. We might think of proportion as a relationship of two (or more) ratios.

Table 2.4 DeKalb County (AL) Property Crime Types by Year

Year	*Prop Crime Total*	*Burglary*	*Larceny*	*MV Theft*
1985	331	162	131	38
1986	380	183	148	49
1987	326	168	122	36
1988	320	191	100	29
1989	189	93	82	14
1990	104	42	45	17
1991	172	99	64	9
1992	192	105	80	7
1993	301	183	114	4
1994	250	83	160	7
1995	406	159	241	6
1996	264	101	158	5
1998	0	0	0	0
2004	271	98	126	47
2006	674	358	210	106
2007	686	187	408	91
2008	535	168	310	57
2009	553	237	285	31
2010	0	0	0	0
2011	908	320	515	73
2012	937	390	471	76
2013	801	333	395	73
2014	810	312	427	71

Using our previous example, we might want to know the number of officers who received complaints in relation to the total number of officers. This inquiry represents a part-to-whole comparison. There are 31 officers, and seven received complaints. In this case, we want a proper fraction, so the numerator will be smaller than the denominator. 7/31 = 0.2258. The proportion of officers that received complaints, then, is 0.2258. That's relatively low, but it would be best for us to reserve judgment until we compare that with other years or other agencies.

Table 2.4 presents much of the same data we looked at before from DeKalb County Sheriff's Office. To be sure, we can report numbers as they fall here. We can say, "Look here. In 2007, there were 686 property crimes known to DeKalb County Sheriff's Office." The natural response is, "So what?" It would not be a disinterested response, but one that sought to pose a question: "what does that mean?" or "why is that relevant?"

In Table 2.4, we can see that in 1991, there were 172 total property crimes, of which 99 were burglaries, 64 were robberies, and 9 were motor vehicle thefts. The number of motor vehicle thefts was low relative to the number of burglaries. Low, perhaps, but we can be more precise. Specifically, we could say that the ratio of burglaries to vehicle thefts, a part-to-part *ratio*, was 99:9, or simplified, 11:1 (99/9 = 11). For every one motor vehicle theft in the DeKalb County Sheriff's awareness, they were aware of 11 burglaries. Expressing the comparison between burglaries and motor vehicle thefts gives us a more palpable sense of the difference between the two in that year.

The next natural question might be to look at the same ratio in previous and subsequent years; to what degree does the ratio of burglaries to motor vehicle thefts change over time? Table 2.5 has a calculated column called "Ratio of Burglaries to Larcenies." The math, of course, is simple: divide the value for burglaries by the value for larcenies for any given

Table 2.5 Property Crimes, DeKalb County (AL) Sheriff's Office, with Ratio

Year	*Property Crime Total*	*Burglary*	*Larceny*	*Ratio of Burglaries to Larcenies*
1985	331	162	131	1.237
1986	380	183	148	1.236
1987	326	168	122	1.377
1988	320	191	100	1.910
1989	189	93	82	1.134
1990	104	42	45	0.933
1991	172	99	64	1.547
1992	192	105	80	1.313
1993	301	183	114	1.605
1994	250	83	160	0.519
1995	406	159	241	0.660
1996	264	101	158	0.639
1998	0	0	0	–
2004	271	98	126	0.778
2006	674	358	210	1.705
2007	686	187	408	0.458
2008	535	168	310	0.542
2009	553	237	285	0.832
2010	0	0	0	–
2011	908	320	515	0.621
2012	937	390	471	0.828
2013	801	333	395	0.843
2014	810	312	427	0.731

year, and you have your ratio. We read the ratio in terms of its part-to-part nature. In other words, we can say that in 1985, for example, there were 1.237 burglaries for every larceny, whereas a decade later, there was only 0.66 burglaries for every larceny. The shape of property crime seems to have shifted in those ten years. Asking precisely why, of course, would be the next step, but that is a project for another time. What is crucial to observe here is that our simple analyses have helped to germinate a more sophisticated inquiry. It bears observing, though, that year-to-year comparisons can be wildly misleading for several reasons (see, e.g., Ashby, 2024), and great care must be taken with such comparisons. The details as to why, though, are for a later time.

We might also ask about the proportion of all property crimes that were burglaries, say, in 2011. In that year, there were 320 burglaries and 908 property crimes. With a little division (320/908), we can see that the proportion of burglaries to all property crimes was 0.352.

Percents

What does 0.352 tell us? I'm so glad you asked! It turns out that there is a kind of proportion (a part-to-whole comparison) that is quite familiar to most people. **Percents** are a special kind of proportion, but they express the comparison made in terms of a standard scale, indicated by the name ("per" meaning "for every" and "cent" meaning 100). Take our burglaries, in 2011, for example. Interpreting the 0.352 outcome, answering the "so what" question is a tad trickier with just this number, but if we expressed it in terms of units of 100 (by multiplying the number 0.352 by 100) we get a **percentage**. It is rather more compelling to say that in 2011, 35.2% of the property crimes were burglaries. We derive this and other percentages by multiplying a proportion by 100:

$$\text{Percent} = \left(\frac{\text{part}}{\text{whole}}\right) \times 100$$

Cumulative Percent

There will be situations when it makes sense to think about the accumulation of events that interest us over time. One way to demonstrate this accumulation is to present the cumulative percent. The reasoning here is that by having a running tally, not just of the cumulative frequency, but also the contextualized tally of the cumulative percent, we can determine, for example, when relatively large increases in accumulation occurred, and when we reached an accumulation milestone, like the midpoint. We will use the cumulative measures a little later on, but for now it serves our needs to invite you to think about something: there are 23 years represented in the table below. If we are watching a cumulative percent increase over the span of these 23 years as we do in Figure 2.1, if the accrual of property crimes is relatively even across years, about when might you expect to see the 50th percentile arrive? Looking at the table again, is your expectation met?

Eureka! The frequency balancing point of the table is rather clearly below the table's midpoint. The implication is that the more recent years have been marked by more property crime than years past; it seems possible that the *rate* of property crimes per year may have increased. Well, OK, a quick look seems to bear that out, but let's also remember that the population has changed over time, too. What we can do is to examine the total amount of property crime, controlling for changes in population. You may be familiar with **crime rates** already from your earlier classes or from the news. Let's make one. Well, let's make a bunch of them. The formula for a crime rate is:

$$\left(\frac{\text{Frequency of incidents}}{\text{Relevant population}}\right) \times \text{base.}$$

The last column in Table 2.6 shows the property crime rate using 10,000 population as the base. Now we are able to begin to ask some more pointed questions about property crime. Wow! Am I right? Do these results indicate that DeKalb County has gotten less safe regarding property crime? Not necessarily. The property crime rate has doubled over the course of the decades shown here, but that *apparent* increase may be an artifact, meaning that it is a spurious outcome. Changes in insurance policies may have incentivized increased reporting over time, or changes in the criminal law or in police reporting practices may have created apparent leaps in rates. In fact, the latter seems increasingly likely when we consider the rather stark increases in crime rates between 2004 and 2006 and between 2009 and 2011. What would be needed now would be institutional memory: what social/economic/legal/procedural changes occurred over the decades represented here. Without any sound sense of things there in DeKalb County, we might conclude that several forces were at work, one of which may include a modest uptick in property offending, but we simply cannot be sure from the data at hand. Crucially, Table 2.6 raises questions (and eyebrows, possibly), but it

Year	Population	Prop Crime Total	Cumulative Frequency	Cumulative Percent
1985	53,922	331	331	3.518
1986	53,642	380	711	7.556
1987	53,927	326	1,037	11.020
1988	54,164	320	1,357	14.421
1989	54,460	189	1,546	16.421
1990	54,763	104	1,650	17.535
1991	55,342	172	1,822	19.362
1992	55,845	192	2,014	21.403
1993	56,484	301	2,315	24.601
1994	57,533	250	2,565	27.258
1995	57,285	406	2,971	31.573
1996	57,374	264	3,235	34.378
1998	58,274	0	3,235	34.378
2004	66,846	271	3,506	37.258
2006	67,829	674	4,180	44.421
2007	68,653	686	4,866	51.711
2008	69,291	535	5,401	57.396
2009	69,781	553	5,954	63.273
2010	71,159	0	5,954	63.273
2011	71,365	908	6,862	72.922
2012	70,944	937	7,799	82.880
2013	70,912	801	8,600	91.392
2014	70,993	810	9,410	100.000

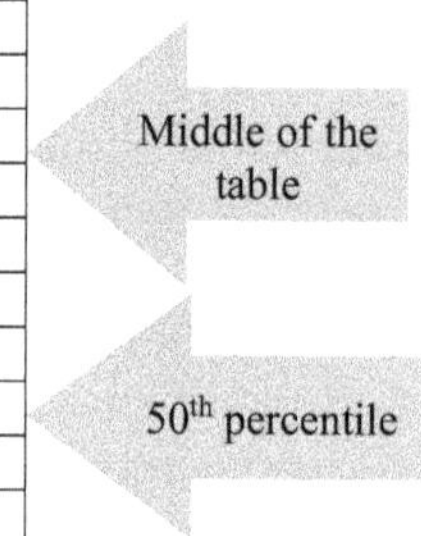

Figure 2.1 Cumulative Frequency and Percents of Property Crimes, DeKalb County, AL

does not provide answers. Crucially, the last column represents the *beginning* of an inquiry, not the *end* of one.

Graphic Presentation of Data

As handy as this frequency table business might be, we have other options for portraying patterns or for deriving context as well. As they say, a picture paints a thousand words. Pictures more plainly tell a story than do tables of numbers, at least for some of us. The aim with visual portrayals is to provide context for scores using color, shape, movement and the like, demonstrating a particular point you seek to make. Graphic presentations tend to have something of the analogy to them; it is not uncommon to see graphic portrayals of data that employ symbols that are directly related to the data they represent.

The realm of graphic presentation of data is truly vast today. Veritable mountains of data have made graphical presentation of data, like data visualization, virtually necessary, and powerful computers have made that visualization possible, even relatively easy. As this text is intended as an introduction, depth and detail will be limited, and principles will be the emphasis.

Pie Charts

We learned from watching the *Men in Black* films and *Twin Peaks* that we do, indeed, need pie. Pie charts are deeply intuitable tools for expressing part–whole relations.

Table 2.6 Property Crime, Population, and Property Crime Rates by Year, DeKalb County, AL

Year	*Population*	*Prop Crime Total*	*Cumulative Frequency*	*Cumulative Percent*	*Prop Crime Rate*
1985	53,922	331	331	3.518	61.385
1986	53,642	380	711	7.556	70.840
1987	53,927	326	1,037	11.020	60.452
1988	54,164	320	1,357	14.421	59.080
1989	54,460	189	1,546	16.421	34.704
1990	54,763	104	1,650	17.535	18.991
1991	55,342	172	1,822	19.362	31.079
1992	55,845	192	2,014	21.403	34.381
1993	56,484	301	2,315	24.601	53.289
1994	57,533	250	2,565	27.258	43.453
1995	57,285	406	2,971	31.573	70.874
1996	57,374	264	3,235	34.378	46.014
1998	58,274	0	3,235	34.378	–
2004	66,846	271	3,506	37.258	40.541
2006	67,829	674	4,180	44.421	99.368
2007	68,653	686	4,866	51.711	99.923
2008	69,291	535	5,401	57.396	77.211
2009	69,781	553	5,954	63.273	79.248
2010	71,159	0	5,954	63.273	–
2011	71,365	908	6,862	72.922	127.233
2012	70,944	937	7,799	82.880	132.076
2013	70,912	801	8,600	91.392	112.957
2014	70,993	810	9,410	100.000	114.096

A pie chart represents parts of a whole by presenting their relative sizes in wedges of a circle. A pie chart makes possible the comparison of the proportion each part represents as members of a whole. Part-to-part comparisons are possible using pie charts, but the specialty of pie charts is the capacity to communicate relative proportions of each part to the whole quickly and clearly. Just a quick glance at Figure 2.2 will tell us that burglaries comprised a little more than half of the property crimes in DeKalb County in 2007, and that motor vehicle theft comprised about an eighth of the property crime there in that same year. The story being told here might be about two characters. First, there's the prevalence of burglaries relative to the total number of property crimes: burglaries make up more than half of the property crimes in that year. The point might be made to emphasize particular policing strategies, like park-and-walks or Operation ID efforts that seek to make burglaries less lucrative, or perhaps to intervene against them directly through an enhanced and visible presence in impacted communities.

Another story one might seek to tell using this pie chart is the relative importance of motor vehicle thefts relative to other property crimes. Again, the point being made might be that motor vehicle thefts were not a pressing issue in 2007 for DeKalb County. Car insurance companies might well want to have a sense of the likelihood of car theft in order to set local insurance rates.

We can figure out how big a given wedge should be by first finding the proportion of the total count for a given wedge. Looking at Figure 2.2, burglaries represent 53% of the total number of property crimes (674, see Table 2.4) for the year. Next, we find the same proportion of 360 degrees; in this case, 53% of 360 is 190.8 degrees, just a little past half of a circle.

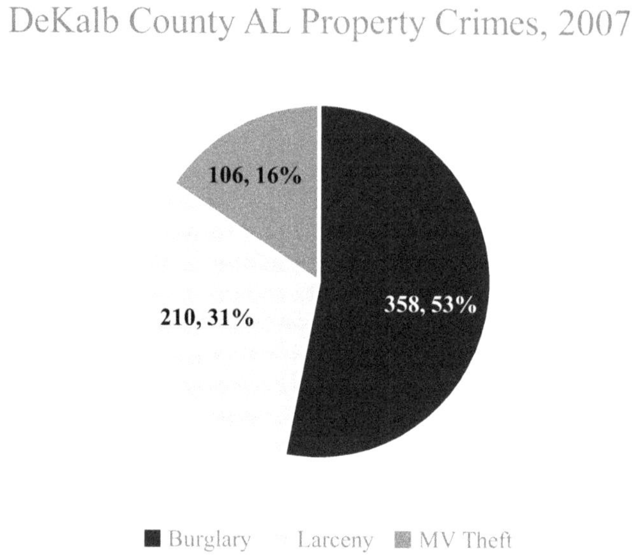

Figure 2.2 Pie Chart of Property Crimes, DeKalb County, AL, 2007

Measured from the vertical line at the top of the pie, we can see that the burglaries wedge does just as we would expect and covers just a tad more than half of the pie (a wedge that had an angle of 180 degrees would be precisely half of the pie, as 180 is one-half of 360, and there are 360 degrees in a circle). The angle for larcenies, which represented 31% of property crimes that year, is $360 \times 0.31 = 111.6$ degrees. And again, as we would expect, the wedge is just a whisker short of one-third of the total pie. The end result, the pie chart, is a clear portrayal of relative amounts of each group within the whole of the property crimes for that year.

One key thing for us to keep in mind is that pie charts, because they presume a part-to-whole relation among the values that correspond to each wedge, assume that there is an additive component to the values. The measures themselves might be of any kind of level of measure (as in the case here – in terms of the level of measure, property crime types are nominal), but the frequencies should add up to a meaningful total (e.g., percentages should add up to 100% of some kind of whole).

Bar Graphs

Whereas a pie chart is a great device for portraying visually part-to-whole relations (or proportions), bar graphs are great for showing part-to-part relations (or ratios). Like the wedges of a pie chart, the columns of a bar graph often present frequency counts of related or comparable events. Although there is something of a part-to-whole message in a bar chart, the main invitation extended by a bar chart is for the reader to compare the size of one bar to another visually. Bar graphs are most useful for presenting frequencies of categorical variables.

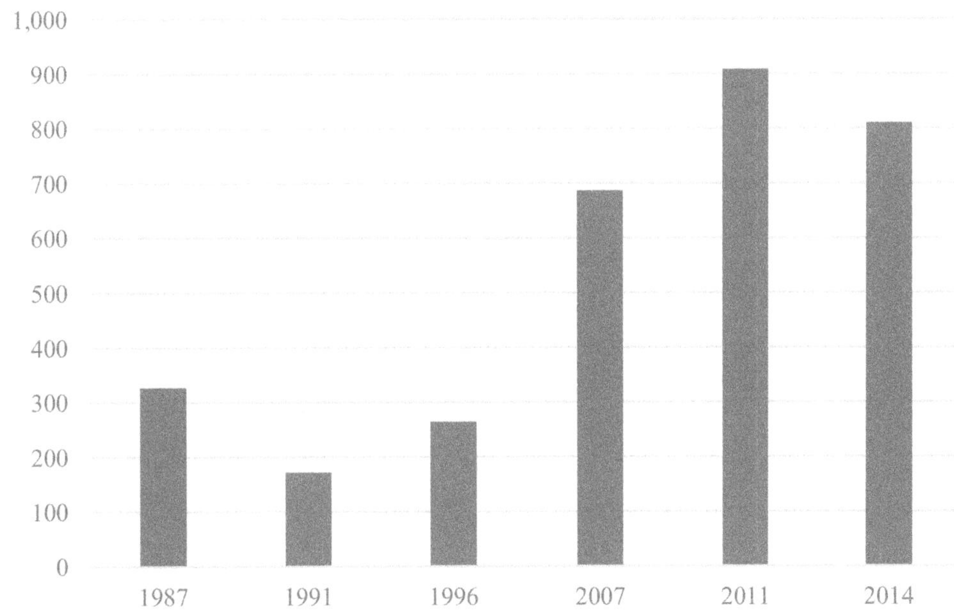

Figure 2.3 Bar Graph of Property Crimes for Selected Years

In terms of their anatomy, bar charts have an *x* axis (the horizontal axis on the bottom) and a *y* axis (the vertical axis on the left). On the *x* axis, there are the values of the variable that interests us. In the case of Figure 2.3, the *x* axis portrays several years. Depending on how the axes are laid out, then, bar graphs like Figure 2.3 can also be fantastic at presenting change in a phenomenon over time.

Notice, though, that these years here are not consecutive, and that the gaps between them are not of a consistent size. These conditions are not optimal, clearly, but again, often we are in a position to use what we have. Notice also that there are spaces between the bars themselves. These spaces matter. In terms of its level of measure, as a variable in this instance, _*year*_ (throughout this book, we will designate all variables as such by italicizing them and adding lines on either side like this) is ordinal – we can order the years, but in this case, we could not subdivide years into months or days. If the time axis was truly linear, meaning that every moment between 1987 and 2014 was represented there, we could present the information a little differently, but we will discuss that in the section on histograms below.

On the *y* axis, we can see a series of numbers that correspond to the number, or frequency, of events. The scaling of this axis can have much to do with what the table seems to say. For example, look at Figure 2.4. The data are precisely the same as in Figure 2.3, but because the way in which the vertical value axis is labeled, the presentation seems to make a different argument, perhaps that property crime is not terribly plentiful. Specifically, instead of going to 1,000 as in Figure 2.3, the axis in Figure 2.4 goes four times as far. As a result,

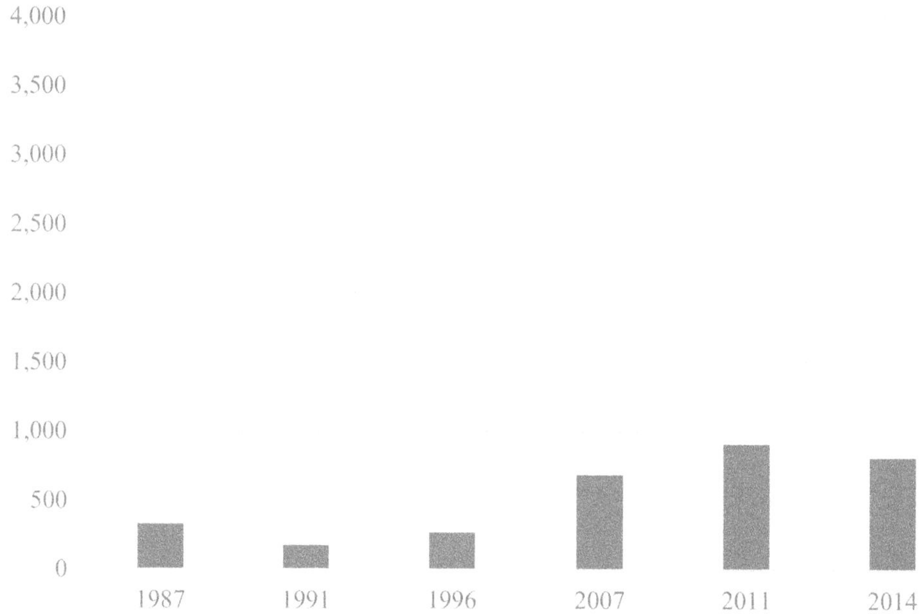

Figure 2.4 Bar Graph of Property Crimes for Selected Years, Modified *y* Axis

the bars look smaller but present the same actual values. More, the relative sizes of the bars are harder to discern when the scale is smooshed like it is in Figure 2.4; the difference between, say, 1996 and 2011 looks smaller in Figure 2.4 than it does in Figure 2.3. Thus, the story we might seek to offer here is that property crime in DeKalb County for these years just was not very high, and also that there has not been much change over the course of the years. Doing so might be an active effort to deceive, of course, to "lie with the truth" and should not be countenanced.

Histograms

Bar graphs like the ones in Figures 2.3 and 2.4 present a count of events (*y* axis) where the *x* axis is categorical. In contrast, Figure 2.5 presents a frequency count in graphic form. The *y* axis is a frequency count, just like in Figure 2.3, but the *x* axis in Figure 2.5 represents values of a *continuous* variable. We can read the values on the *x* axis, in this case, scores on a scale that seeks to measure citizens' fear of crime, as if they sat on a number line. Notice that there is no space between the several bars. Frequency histograms, like this one, present a frequency count of events that are, from a level-of-measure perspective, continuous (that is, interval or ratio). The story told in Figure 2.5 is in relation to the larger shape of outcomes seen in the data, and not necessarily about the particular value of one part compared with another or to the whole.

And note finally the shape of the bar pattern in Figure 2.5. It has a distinct bell-like shape, but it is not precisely a bell; it is more of a bell in need of a haircut. Figure 2.5 represents the

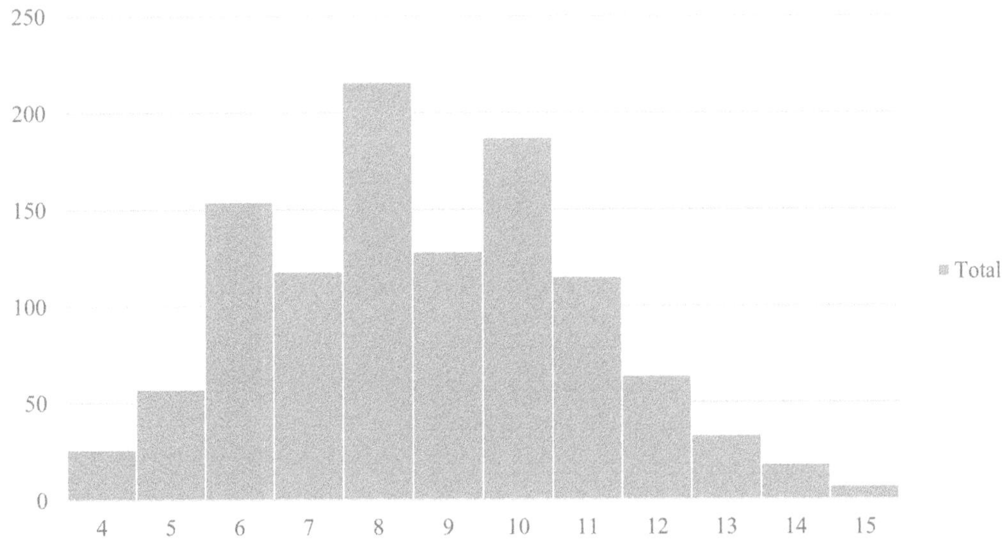

Figure 2.5 Histogram, Fear of Crime in Hypothetical City

first time we will pay attention to this shape in a distribution; it is enough for the moment merely to notice it. More on this is soon to come.

Continuous Variables and Theoretical Continuity

And just to be clear, "continuous" does not mean that they are *actually* occurring perpetually, but instead that they are numeric in nature. Recall that the term "continuous" is in relation to interval and ratio measures. That means that they are numeric (the numbers actually specify numeric value) and that they are subdividable (see Figure 1.2). Because they are numeric, they sit on a number line that, in theory, continues forever in either direction, and also "into" the bar through infinitely increased precision, at least theoretically. Continuous measures, then, are not *actually* (empirically) perpetual, but are *theoretically* boundless. Phenomena like human age or height or number of crimes charged on a docket have physical and practical limits, and precision, well, that gets into another whole issue to come later.

Polygons

There are a host of different graphical techniques that present data using polygons, like Figure 2.6 does. In Figure 2.6, the data from Table 2.4 are presented. Each polygon represents the amount of each property crime in the year indicated on the x axis. The story told in Figure 2.6 is about the larger patterning of outcomes in terms of their relative frequency, but the lines themselves tend to emphasize the change points in the continuity of relative frequencies. Often, as is the case in Figure 2.6, polygons are used for expressing change over time. Here, the x axis represents the movement of time over the course of several years.

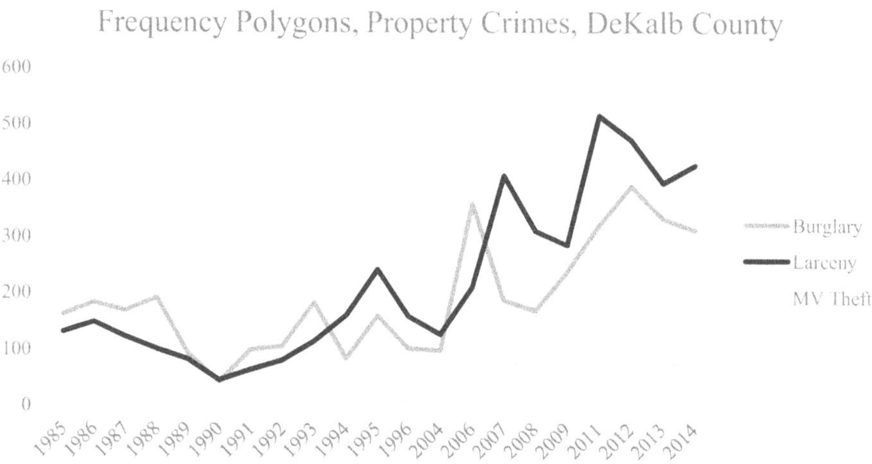

Figure 2.6 Trend Lines for Property Crimes, DeKalb County, AL.

Very quickly, by following the lines on the page, we can see something of a footrace between burglaries and larcenies in DeKalb County; although burglaries were more common in the 1980s, the footrace between burglaries and larcenies, at least as far as 2014 is concerned, seems to have been "won" by larcenies. The number of motor vehicle thefts had a peak in the early 2000s but seems to have leveled off since. Polygons can be seen as implying a pattern of **trend** as well; burglaries and motor vehicle thefts appear to be trending downward slightly or showing the possibility of being on a vector toward becoming relatively fewer, whereas larcenies appear to be trending up. These trend patterns might either be a reason to celebrate, say, if DeKalb County had been engaged in an anti-burglary programming of some kind around 2006. It may also be something of a call to action if the emphasis is on the upward-trending larceny "problem." It may *also* be an artifact of changes in law, policy, reporting or recordkeeping, and thus represent nothing whatsoever. In any event, the meaning of the figure is the story you seek to understand.

Distributions, Variability and Uncertainty

Let's take a second to look a little more closely at these last two graphical tools: histograms and polygons. They both do a great job of giving a useful sense of the distribution of data in continuous (interval, ratio) variables, but we might still be craving additional context to support the drawing of more definitive conclusions. As we have seen, these graphical tools can raise more questions than they tend to answer.

You may have noticed the shape of the histogram we used as an example. This shape is important. It looks like a bell, sort of. For reasons we will explain later in Chapter 5, we can use a **theoretical distribution** to study real-world, or **empirical distributions** like this one. Put another way, we use a mathematically derived theoretical template to provide context for comparison with our empirical continuous variables. We call the first of many theoretical distributions the normal (or Gaussian, or bell) curve. This mathematically derived curve often serves to provide some measure of context for comparisons that can support conclusions. The curve itself comes from probability theory, something we will also explain in Chapter 5. It helps us to know right now, though, that we have two different but related curve things: empirical data which represent real, measured data, shown in histograms and such, and theoretical curves, like the normal curve, which come from theory rooted in probability experiments (like repeatedly flipping a coin – we'll get to that, too) and are used to provide context for those empirical measurements. In this text, the bell curve in a few forms will be the primary means for providing context related to expected outcomes, but this method comes with some problems and is not the only method available to you.

You see, in the world of quantitative empirical data, by that I mean data we derive from measuring things in the world that interest us, the distributional shape of values in real sets of scores do not always (or even often) look like the theoretical bell curve. Sometimes real data are piled to one side or another (which we call skew), sometimes due to a few very high or very low scores. Sometimes they look nothing at all like a bell curve, and in terms of the phenomenon itself, shouldn't be expected to be. This is key, of course: the bell curve provides a comparison to expectation, but it does not necessarily provide a template for assumed phenomenon. If what we expect a variable to be like is something other than a bell curve, it does us no good to make such a comparison. Or at least it would seem so. Again, that's also a thing for later conversations.

Suggestions for Selecting and Using Presentation Options

1 Know your data and know your variables aside from the ones you are using. Where did the data come from? Who created it? What was the purpose behind the data?
2 Even at this early stage, you are telling a story. Do your level best to start your story with some sense of how the whole story will unfold.
3 Examine in detail the variables you plan to use. Right now, of course, we are thinking about univariate techniques. Know the level of measurement for the variables you seek to present.
4 The story you want to tell will be a thing that will emerge from the data. Don't make the data say something they don't, and do your best not to occlude interesting things you find.
5 Don't rely on fancy things; often the most impactful graphical presentation of data is simple, elegant and to-the-point. Be elegant, parsimonious. Don't use five figures when one will do.

Main Take-Aways

- For our purposes in this book, analysis can be understood in terms of comparisons utilizing some kind of context for framing comparisons.
- Graphical presentation of data, by its very nature, is a process of analysis or data reduction.
- Expressions like proportions, ratios or percentages are also kinds of data analysis.
- It is exceedingly easy to lie with statistics, and graphical presentations of data are a powerful tool for lying. Such lying, of course, is unconscionable, and you'll not do it. Right?

Key Terms (in order of appearance)

frequency table
context
cumulative percent
bar graphs
reduced (i.e., data reduction)
comparison
normal forms
proportion
theoretical distribution
research question
ratios
crime rates
histograms
empirical distribution
levels of measurement
imputing [i.e., imputation]
cumulative frequency
empirical distribution
cumulative percent
bar graphs
interval
percents/percentage
pie charts
polygons

Chapter 3

Centrality

Although graphical presentations and frequency tables will come in handy, these things may not be handy enough for the chief or her command staff. The "typical case" question might not have seemed so intuitive at first, but almost certainly you have dealt with a similar question in some form or another sometime in your life. **Typicality** is a way of talking about a set of scores in a very concise way. What is the number of years of service for the typical officer in your department? Sure, there are some with decades of service, and some who are just starting out, but what is the most representative service length among officers in the department?

Recall in the last chapter how we talked about empirical and theoretical curves, like the bell curve or like our example here of a continuous variable (years of service)? In terms of typicality, for either instance, it is perhaps most useful to think about typicality in terms of where these distributions would balance if we could place them on a point.

Not all variables are continuous, though, as we already know. There are several ways of measuring centrality, and it turns out that which of these measures we use is dependent upon the level of measure of the variable we want to describe.

Just a note: in our first look at some of the measures of centrality and dispersion, it helps us to think about which of these measures have use beyond the question they are designed to answer. As we will see, some measures are less useful beyond themselves, whereas some other measures are used in a number of other statistical recipes.

The Mode

Our first option is the **mode**. The mode is embarrassingly simple: it is the most frequent score in a set of scores. That's all there is to tell you, really, at this point.

Modes are particularly useful for helping to portray categorical, that is, nominal or ordinal, data. It can be used for continuous data, too, but there can be some bugs that crop up when we do. With a nominal variable, like, say, religious preference of jail internees, as the variable is nominal (no one religion is more religion than another), the most direct way of talking about the typical internee is to find the modal, or most common, answer to the question, "What is your religion?" In some places in the United States, we might find that the modal religion among internees in a jail is "Christian" mostly because in many places in the United States, there is a relatively high number of people who identify as "Christian."

Modes are useful for presenting nominal or ordinal data. Modes do not make use of all the information available in a set of scores, but in some ways that's not a horrible thing.

DOI: 10.4324/9781032638591-3

Modes do not get yanked up or down by extreme scores (we will talk about extreme scores later in the book). At the same time, when used with ordinal data, a mode may not represent the middle, or center of gravity, of a set of scores. It is possible for the modal value to be near the high or the low end of the range. With continuous variables, difficulties are only amplified; often with a continuous variable, each score occurs only once (depending on the amount of precision being employed), so the modal value might occur only two times out of a wide range of scores. The mode could shift with only the smallest change in value, there could be multiple modes, or no mode at all. As level of measure moves from categorical to continuous, the relative usefulness of the mode tends to decrease, although there could be times when the most frequent score of the values of a continuous variable might well be a thing worth reporting.

Mode strengths:

- Robust against effects from extreme scores
- Powerful measure of centrality for nominal and ordinal variables
- It may be the only real choice for nominal variables
- Versatile in that multiple meaningful modes can exist.
- It enjoys rather *occasional* use beyond the thing itself.

Mode weaknesses:

- Can be arbitrary with continuous variables
- Does not leverage much information from the variable itself
- Potentially unhelpful in that no mode or many modes may exist, and they may give little useful sense of a mid-point or of typicality.

The Median

The second measure of centrality is called the **median.** The median in a road is an area that lies in the middle between the lanes of traffic on either side. The median of a set of scores is the middle score. Of course, to find the middle score, *we need to place the scores in value order* (and that means that the variable must be orderable, so it must be at least ordinal, if not interval or ratio). Once we order the scores, we can find the median by checking off values from each end until we have either one or two left. If we have one, that score is the median (Wooo!). If we have two, we add them together and divide by two (also Wooo!).

Sometimes there are too many scores to line them up and notch them off. Another way to find the median is to identify its *position* (once the values have been placed in order) using the formula:

$$p_{\text{Md}} = \frac{n+1}{2}$$

where p_{Md} is the *position* of the modal value, and n stands for the number of data points or scores. Remember, though: this formula *does not give you the median itself*. It only helps you find where the median is. It gives you the median's *address*. It arranges the introduction but doesn't actually make it.

Median strengths:

- It is robust against influence of extreme scores.
- It is a solid expression of "middle-ness" and not simply the most commonly occurring value.
- It has some uses beyond the measure itself.

Median weaknesses:

- Although it leverages more information than the mode does, it does not contain in it as much information about the distribution as we might like.
- Although it has a number of uses beyond itself, it is not widely relied upon in other statistical recipes.

The Mean

Last but surely not least, there's the **mean**. The mean is, like the mode and the median, very intuitive in terms of how it works. The formula for the arithmetic mean ($\bar{x}$) is:

$$\bar{x} = \frac{\sum x}{n}.$$

We add up all the scores and simply divide by the number of scores. We are taking all the information represented by the set of scores, mashing it together into a single pile and then dividing it into *equal units*. The size of the equal units indicates the balancing point of the distribution.

Mean strengths:

- Utilizes all available information from the variable.
- Is used widely beyond itself in other statistical recipes.

Mean weaknesses:

- Subject to being influenced by extreme scores.
- Arithmetic mean is "volatile" when "n" (**that is, the number of scores**) is low.

Let's calculate the mean number of years of service for the officers in our department.

$$\bar{x} = \frac{\sum x}{n}.$$

So the first thing we need is to sum (add) all of the values of x: 10.44 + 22.92 + 15.36, and so on.

Adding them all together, we get 355.58. That number itself is an ingredient, not so much one that has much flavor on its own. In terms of n, we have 30 officers, so $n = 30$. We can just slap in the numbers and roll:

$$\bar{x} = \frac{355.58}{30}.$$

Table 3.1 Years of Service of Sworn Officers in Your Department

Officer	*Years of Service*
Roberts	10.44
Van Slyke	22.92
Lenning	15.36
Leneker	20.80
Adkinson	20.07
Vollum	20.66
Garland	19.37
Brooker	21.34
D'arcangelis	6.75
Halpin	4.06
Sklar	3.12
Sparks	9.22
Barile	4.89
Hicks	7.57
Wright	4.15
Brightman	8.29
Barlow	5.81
Vandewerker	5.13
Lyons	3.00
Powell	7.67
Hayes	2.88
Adelman	11.46
Jonker	12.34
Bennett	14.14
Klemme	21.63
Collins	13.17
Fiebig	15.50
Cannady	13.95
Smith	10.51
Kretser	19.38

And the answer is...

$\bar{x} = 11.85$ years.

What this means, then, is that the typical officer has served nearly 12 years in this department. We will put this insight to work for us more fully in subsequent chapters, but for the moment, take this information and look back at Table 3.1. Which officers have the "typical" amount of experience? Which ones have more? Who might you want to learn from, and why? Who would you want coming to a call for service that you placed? Why do you say so?

While we're at it, let's do a median as well for the same data. First, of course, we must place them in order:

2.88, 3.00, 3.12, 4.06, 4.15, 4.89, 5.13, 5.81, 6.75, 7.57, 7.67, 8.29, 9.22, 10.44, 10.51, 11.46, 12.34, 13.17, 13.95, 14.14, 15.36, 15.50, 19.37, 19.38, 20.07, 20.66, 20.8, 21.34, 21.63, 22.92

First, let's find the median's position.

$$p_{\text{Md}} = \frac{n+1}{2}.$$

In this instance, we know there are 30 scores. That means $n = 30$. Let's do the thing:

$$p_{\text{Md}} = \frac{n+1}{2}$$

$$p_{\text{Md}} = \frac{30+1}{2}$$

$$p_{\text{Md}} = \frac{31}{2}$$

$$p_{\text{Md}} = 15.5.$$

Well, so the result is a decimal. Odd, right? Nope. That's because the median position is *between* two scores. That there isn't a single median shouldn't surprise, because there are 30 scores (we would also say that $n = 30$), which is an even number. If n had been odd, there would be one middle score.

Say we wanted to find the median itself. If you notch one score off each end until they meet in the middle, you'll find again that there's no one middle score. 10.51 and 11.46 are the two middle scores. Summed, they total 21.97. In this circumstance, we would find the mean of the two middle scores:

$$\bar{x} = \frac{\sum x}{n}$$

$$\bar{x} = \frac{21.97}{2}$$

$$\bar{x} = 10.985.$$

Our median, then, is 10.985. Now the arithmetic mean was 11.85 years, a little higher than the median, but defensibly close. Did we make a mistake? No, these two measures of centrality approach the question differently as you already know. In fact, as we will see in the next chapter, the different approaches to centrality used by the mean and the median will come in handy for yet other questions. We'll soon see that the difference between mean and median or mean and mode can tell us important things about our data.

Weighted Mean

On occasion, there arises the need to treat some scores in a range a little differently than other scores. You and I are well familiar with one such circumstance: grades in a class. Say, for example, a professor wants a particular assignment to count for more than others. Maybe that assignment is a test, and she wants it to matter more than homework tasks. Well, she might attach a weight to that test such that it counts for more than those homework assignments do. As a result, that test will have greater influence on the mean than will any one of those homework tasks. Perhaps these are the tasks for the class: five homework assignments, each worth 10% of the total class grade; one midterm worth 20%, and one final worth 30%.

Together, of course, that's 100% of the points. That final grade will have the most tug to it relative to the mean that results. Let's say a student gets the following grades:

Homework 1: 82
Homework 2: 79
Homework 3: 83
Homework 4: 81
Homework 5: 78
Midterm: 81
Final: 94.

For this student, each grade will be multiplied by its respective weight. After, we sum the weights themselves, and use that as the denominator. The formula looks like this:

$$\bar{x}_w = \frac{\sum(xw)}{\sum w}$$

where
x = each score
w = the weight given each score
(xw) = is the product of the score times the weight
Σw = is the sum of the weights.

That means, then, that we would have the following (noting that the percentages have been changed into decimals for convenience here):

Homework 1: 82 × 0.1 = 8.2
Homework 2: 79 × 0.1 = 7.9
Homework 3: 83 × 0.1 = 8.3
Homework 4: 81 × 0.1 = 8.1
Homework 5: 78 × 0.1 = 7.8
Midterm: 81 × 0.2 = 16.2
Final: 97 × 0.3 = 29.1.

Summed, we have 85.6. Our last step, summing the weights, is easy here because we know they equal 1.0 (they're decimal expressions of percentages). 85.6/1 = 85.6. That final brought up the average enough for a bump in the final grade relative to how the homework tasks had been going.

Other Kinds of Means: The Pythagorean Family

The arithmetic mean is the version of the mean with which most people are familiar, but it is merely the most famous among the three **Pythagorean means**. There are two other kinds of means we should mention here, mostly because they are useful in particular situations. It is important to point out that although the geometric mean and the harmonic mean are well-suited to specific questions, the arithmetic mean is also the best choice at times, depending on what is sought. No hard and fast rule guides which measure is always better. You as the researcher need to determine that according to conditions and intention.

The **geometric mean** is another way of identifying typicality, but instead of adding numbers together, it multiplies them, and then takes the *n*th root to find the value. Why would we do this? Because not all things are additive in nature. Some are multiplicative in nature, so a geometric mean can be more precise. A geometric mean also tends to be more robust than an arithmetic mean is against the influence of extreme scores. A geometric mean, then, is a good choice when a variable has some very high or very low scores. Calculating the average growth of a population or the average growth of a financial investment are also sound uses for geometric means. Here is the formula for calculating a geometric mean:

$$\bar{x}_g = \sqrt[n]{x_1 x_2 x_3 \ldots x_n}$$

Another interesting benefit of using the geometric mean is that it works across different scales. Say you are looking to compare the average ratings for your police department to similar ratings for another department according to two different surveys. One hitch is that the two surveys used different scoring systems for their ratings: one instrument asked respondents to rate department attributes on a scale of 1–5, whereas another instrument used a scale of 1–20. A geometric mean can be used to average the ratings and compare the two departments without the average being influenced by the scales of the two rating systems. When we combine scales like this, though, the geometric mean is no longer expressed in a unit related to the information you put in front of it. For basic decision purposes, this may well be just fine. As always, care and discernment are called for when choosing the right measure for the question.

Say our municipality has grown in population over the past several years, and so the chief wants to hire more cops. She needs to present to the city council information about population growth to further her argument about the need to bring on more officers. The city's population growth has been big, but not consistent:

If we calculated an arithmetic mean for percent growth, we would get:

$$\bar{x} = \frac{2.1 + 4.9 + 8.2 + 11.6 + 6.7 + 5.5}{6} = \frac{39}{6} = 6.5$$

A geometric mean for the same percentages would be:

$$\bar{x}_g = \sqrt[6]{2.1 \times 4.9 \times 8.2 \times 11.6 \times 6.7 \times 5.5} = \bar{x}_g = \sqrt[6]{36{,}068.22} = 5.748$$

Cooking a geometric mean involves two basic steps. First, find the product of the scores. Multiply the several scores under the bracket from left to right. Second, find the *n*th root of the product of those scores. Poof! You're done.

To take the *n*th root of a number, just raise that number to the power of $\frac{1}{n}$. Or if you prefer, use $\frac{1}{n}$ as a decimal. In this case, $\frac{1}{6}$ becomes 0.16667. Why is the 6 in the denominator? Because there are six scores. Raise 36,068.22 to the power of 0.16667 and you get 5.748.

The geometric mean is pretty cool and all and is acutely useful for instances where it is most appropriate, but it has something of a blind spot. The geometric mean is just the thing for speaking to the typicality of a set of growth percentages. The thing is, though, the geometric mean is not workable mathematically with a negative number in the frame, and negative numbers are meaningful possibility with growth percentages. We should not abandon the geometric mean logic, so we need to find some kind of a workaround.

It turns out that the workaround is pretty simple. Instead of using the percentages themselves, we can use the multiplier that expresses the same value. It involves a few steps, and in the end, it has to be reversed, but it isn't tricky at all. Actually, the process itself has use outside the current context. If you wanted to figure out a proper tip for a nice meal, we can use this method. Say, for example, we enjoy a nice lunch, spending US$47.84 (I had to try the soup). A proper tip of 20% can be found by:

$$\$47.84 \times 0.2 = \$9.57$$

The 0.2 is our 20%, of course. We would then add US$9.57 (which is 20% of the meal cost) to the US$47.84 to get the total of US$57.41. Easy enough, but what if I could show you how to shorten the process to one simple step? Very well then. Instead of multiplying the total to get the tip *and then* adding them together, multiply US$47.84 by 1.2. The 1.0 preserves the US$47.84 value, and in effect, combines the two steps into one.

$$\$47.84 \times 1.2 = \$57.41$$

Slick, right? Well, finding the 1.2 business for a meal is easy, because we're deciding on the percentage. We're picking the number we're going to use and just tacking a 1.0 on the front. How does this help us with the negative percent growth problem?

It helps us because it gives us a number that can work in our geometric mean. The first step is to take the percentage and divide it by 100. The second step is to add 1.0 to the result, just like the tip example. That's it. Say that, compared with the year before, a city had a 2.3% increase. 2.3/100 = 0.023. That's step one. The second step is even easier: 0.023 + 1 = 1.023. That 1.023, which is the number you'd multiply the previous year's percent growth by in order to find the new one, contains the information we need. Now, if we have that pesky negative growth percentage for a year, say a city basically shrinks by 1%:

$$-1/100 = -0.01$$

$$-0.01 + 1 = 0.99$$

So 0.99 becomes the multiplier we would use in a geometric mean. It is geometric-mean-friendly and it contains the information we need. To find that geometric mean, then, we calculate the geometric mean just as we did above, but instead of using the percentages, we would find the corresponding multipliers, perform the geometric mean on them, and then turn that number back into a percent growth. Once we have the result, we simply reverse the process.

Say we have the following growth percentages: 3.2, 1.1, 0.02, −1.3, −0.04. Let's find the geometric mean for them. First, of course, we need to make the multiplier values:

$$3.2/100 = 0.032 \rightarrow 0.032 + 1 = \mathbf{1.032}$$

$$1.1/100 = 0.011 \rightarrow 0.011 + 1 = \mathbf{1.011}$$

$$0.02/100 = 0.0002 \rightarrow 0.0002 + 1 = \mathbf{1.0002}$$

$$-1.3/100 = -0.013 \rightarrow -0.013 + 1 = \mathbf{0.987}$$

$$-0.04/100 = -0.0004 \rightarrow -0.0004 + 1 = \mathbf{0.9996}$$

Not too bad, right? OK, now we can get to the business of the geometric mean itself. Note that there are five scores, so $n = 5$:

$$\bar{x}_g = \sqrt[n]{x_1 x_2 x_3 \ldots x_n}$$

$$\bar{x}_g = \sqrt[5]{1.032 \times 1.011 \times 1.0002 \times 0.987 \times 0.9996}$$

$$\bar{x}_g = \sqrt[5]{1.02958238}$$

$$\bar{x}_g = 1.00584768$$

Recall that we can get that nth root by raising our product to the power of $\frac{1}{n}$, which in this case is $\frac{1}{5}$ or 0.2. OK, but we're not quite done. The final step gives us our answer. We need to take back the 1.0, and then move the decimal two slots by multiplying by 100.

$$1.00584768 - 1 = 0.00584768 \rightarrow 0.00584768 \times 100 = 0.584768$$

In this example, then, the city with two years of negative growth had a five-year geometric mean growth of just more than half of 1% per year, roughly 0.58%.

So why use the geometric mean and not the arithmetic mean in this instance? Well, look back at Table 3.2. You will see that in 2015, there was rather considerable growth in the population of the city. An arithmetic mean would be tugged up by a score that varied significantly from the rest. Notice also that 2012 was rather slow, growth-wise. The arithmetic mean, 6.5 frames 2012 as potentially problematic from a growth perspective; something might well have gone wrong that year, and we might be driven to find some cause. The geometric mean reduces the seeming problem of 2012. Remember: the purpose of a measure of central tendency is to express the idea of *typicality*. If neither 2012 nor 2015 were typical in the fullest sense, it would serve us ill if they had an outsized influence on our understanding of typicality.

The **harmonic mean** is the third member of the Pythagorean mean family. It is also multiplicative, sort of. Classically, harmonic means are best known for being the proper measure for finding the mean speed or rate of travel over a distance. The formula for the harmonic mean is:

$$H = \frac{n}{\frac{1}{x_1} + \frac{1}{x_2} + \frac{1}{x_3} \ldots \frac{1}{x_n}}.$$

Table 3.2 Population Growth for Hypothetical City

Year	*Percent Growth*	*Population on 31 December*
2012	2.1	89,031
2013	4.9	93,393
2014	8.2	101,052
2015	11.6	112,774
2016	6.7	120,330
2017	5.5	126,948

Table 3.3 Officer Speeds on a Measured Length of Patrolled Highway Three Times of the Day

Officer	*Date*	*3:00 a.m.*	*8:00 a.m.*	*11:00 a.m.*
Adkinson	1 June	64 mph	38 mph	54 mph
Vollum	2 June	71 mph	35 mph	61 mph
Garland	3 June	74 mph	18 mph	52 mph
Brooker	4 June	60 mph	11 mph	48 mph
VanDiver	5 June	66 mph	24 mph	50 mph

The harmonic mean is the reciprocal of the sum of the reciprocals of the several scores. OK. First, a reciprocal of a number is 1 over that number, or $\frac{1}{n}$. We just did that above. To make a reciprocal of a number, we simply make it the denominator of a fraction, and put a 1 as the numerator. It is also called the inverse of a number.

On each of five consecutive days, five different officers drove the same two-mile length of southbound measured highway as part of their duty shifts (they didn't know their speed was being measured). Speed was measured for each of these trips; on June 1, Officer Adkinson drove 64 miles per hour for the entire length of road around 3:00 a.m. He returned heading the same direction again around 8:00 a.m. and drove 38 miles per hour this time, mostly because of traffic. He drove back north via side streets over the course of his patrol, and near the end of his shift, he drove the same direction down the same length of highway around 11:00 a.m., going 54 mph the whole way. Officer Vollum did much the same thing on the following day, and so forth.

We might be interested in the average, or typical, speed of each officer (e.g., if driver safety is a concern) or perhaps we might also wonder about the impact of morning commuters on the use of a southbound interstate through the jurisdiction (e.g., for response efficiency). Let us examine both of these questions.

First, let us examine the harmonic mean of Officer Adkinson as he drove this same length of road three times the same day.

Again, our formula for a harmonic mean is:

$$H = \frac{n}{\frac{1}{x_1} + \frac{1}{x_2} + \frac{1}{x_3} \ldots \frac{1}{x_n}}.$$

In the case of Officer Adkinson, we can plug in numbers:

$$H = \frac{3}{\frac{1}{64} + \frac{1}{38} + \frac{1}{54}}.$$

First, we do the business on the bottom. $\frac{1}{64} = 0.015625$. $\frac{1}{38} = 0.026316$. $\frac{1}{54} = 0.018519$.

$$H = \frac{3}{0.015625 + 0.026316 + 0.018519}.$$

Then we do the addition:

$$H = \frac{3}{0.06046}.$$

Then we solve.

$$H = 49.62 \text{ mph.}$$

Officer Adkinson averaged 49.6 miles per hour that day on that stretch of highway. For the sake of comparison, let's look at the same numbers slotted into an *arithmetic mean*:

$$\bar{x} = \frac{64 + 38 + 54}{3} = \frac{156}{3} = 52 \text{ mph.}$$

The harmonic mean figures in the ratio component of the measure, that it is a *miles-per-hour* metric. An arithmetic mean is rendered imprecise by the fact that what we seek is not a simple number, but is a ratio, an expression of *two* measures in one (velocity expressed in terms of **miles per hour**).

Now let's look at the 3:00 a.m. traverses of that piece of road by our five officers.

$$H = \frac{5}{\frac{1}{64} + \frac{1}{71} + \frac{1}{74} + \frac{1}{60} + \frac{1}{66}}.$$

First, we calculate the decimal values for the reciprocals in the denominator.

$$H = \frac{5}{0.015625 + 0.01408 + 0.01351 + 0.016667 + 0.015151}.$$

Then, we add the decimals in the denominator.

$$H = \frac{5}{0.075033}.$$

Finally, the coup de gras:

$$H = 66.64 \text{ mph.}$$

What conclusions might we draw from what we have seen regarding the differences in speed, or about the officer-wise differences we see? Do we have enough information to make any definitive decisions?

Summary

Our measures of centrality give us straightforward and direct ways of expressing the typicality of a distribution of scores, whether those scores be nominal, ordinal, interval or ratio. These expressions of typicality will, as we will see shortly, give us something of a benchmark for our other essential description of our distributions: dispersion.

What Did We Accomplish?

We became acquainted with several measures of centrality, or typicality, as well as how best to use them.

Main Take-Aways

- For our purposes in this book, analysis can be understood in terms of comparisons utilizing some kind of context for framing comparisons.

- Graphical presentation of data, by its very nature, is a process of analysis or data reduction.
- Expressions like proportions, ratios or percentages are also kinds of data analysis.
- It is exceedingly easy to lie with statistics, and graphical presentations of data are a powerful tool for lying. Such lying, of course, is unconscionable, and you'll not do it. Right?

Key Terms (in order of appearance)

typicality	mean	geometric mean
mode	weighted mean	harmonic mean
median	pythagorean means	

Chapter 4

Variation and Dispersion

If you want to specify an address in an urban setting, often we express that address in terms of the street it is on, but also a cross-street in the grid of city blocks. The address for, say, Chelsea Market in Manhattan, New York City is 75 9th Avenue. It is more helpful to many folks, particularly for taxi drivers, though, to describe it as being on 9th Avenue at 15th Street.

If you live in Manhattan, you might well know precisely where Chelsea Market is. If you don't have a good idea of where it is, a cross-street has practical value for finding the place in space. Metric variables operate in much the same way: knowing that Chelsea Market is on 9th Avenue in Manhattan is helpful but knowing that it is near 15th Street and not by 60th Street on 9th Avenue, well, that makes quite a bit of difference. Just as important as typicality for describing a variable is the amount of variability present in that variable. If the main address is analogous to variable typicality, we might think of variability as being akin to the cross-street of an address. One set of scores can have the same mean as another set of scores but have a very different amount of variability.

Knowing where Chelsea Market is, of course, only gets you there. The cross-street is just its location. It does not give you the merest sense of all the things on offer inside. It is important not to equate typicality and variability with the rich nuance within a variable. Typicality and variability are the height-and-weight descriptions of a variable. They will become useful as references for us in our analyses, but they themselves do not contain the world entire.

Measures of Variability

Just like with measures of centrality or typicality, the level of measurement of the variable matters greatly regarding the best choices for the expression of variability.

Range

Perhaps a first step to seeing the nature of variability in a variable is to consider the **range**. The range of a variable is, simply put, the distance between the smallest and the largest valid values in the set of scores. To calculate the range for the years of service variable, we simply subtract the smallest score from the largest. Look back at Table 3.1. You'll see that Officer Brooker had 21.34 years of service and Officer Lyons had only 3.00 years. These two officers represent the highest and lowest values for the variable "years of service." The range of scores for that variable is $21.34 - 3.00 = 18.34$.

DOI: 10.4324/9781032638591-4

Most often, the range can be used for any variable (nominal, ordinal, interval or ratio), but perhaps more robustly so for variables that are orderable in some fashion. The example above from Table 3.1 uses a variable that is interval. It is also usable for nominal variables, but in such instances, it is little more than a report of all values that have at least one valid outcome.

Variation Ratio

In order to talk with greater precision about the amount of variability we find in a categorical variable, we can think about it in terms of a comparison. Recall how in Chapter 1 we talked about how statistics are often derived through making some kind of comparison? The variation ratio is a good example of this. The formula for deriving a variation ratio is:

$$VR = 1 - \frac{f_{\text{modal}}}{n}$$

Basically, the variation ratio expresses the proportion of cases in the set of scores that *do not* fall into the modal category (that's why we do the 1-minus bit). As is the case with any proportion, we can say it as a percent by multiplying it by 100. Like the range, the variation ratio is not used much in other statistical operations. Let's cook one.

Say we are looking at religious preferences for internees in a jail, like from before. Let's say that our internees look like this (no, it is unlikely, but it works to serve the point):

What we want to do is to find the proportion of internees who *don't* belong to the modal category of religion. The total number of internees at this point is $n=59$. The mode here, clearly, is Christianity ($n=25$). We can roll, then:

$$VR = 1 - \frac{25}{59}$$

Then we do the division:

$$VR = 1 - 0.4237$$

And then the 1-business:

$$VR = 0.5763$$

If we multiply the VR by 100, we get a percentage, specifically, the percentage of internees who are not of the modal category. In this case, it would be 57.63%. We might expect, then, that if Christian services are offered, a majority of internees are unlikely to take advantage of them.

Mean Absolute Deviation

For continuous variables, we can do quite a bit more with them to get precise and useful measures of their variability. One way of thinking about variability in a continuous variable is to consider each score's distance from the mean. The variation ratio conceptualizes variability as the proportion of cases that are not in the modal category as we just saw. It is comparing the center of gravity (the modal category) to everything else. It tells you the proportion of scores that are *not* typical. The example in Table 4.1 shows that the majority

Table 4.1 Religious Preferences among Hypothetical Jail Internees

Religious Preference	*Frequency*
Bahá'í	4
Buddhism	6
Christianity	25
Judaism	12
Muslim	8
Shinto	3
Zoroastrian	1

of cases are not, in this sense, typical. Well, with continuous variables, we can take that logic rather further. We can compare *each score* to the typical case. Instead of the mode, though, we use the mean. As we are making a comparison, we use subtraction. Let's take an example, revisiting a familiar example.

Looking back at Table 4.2, we find the mean years of service was 11.85 years. Let's see what we get when we compare each score to the mean using subtraction.

OK, so we can have some sense of how much each score varies from the mean. Now it makes sense to talk about the *typical amount of variability around the mean.* Let's look back for just a second at the formula for arithmetic mean:

$$\bar{x} = \frac{\sum x}{n}$$

In this formula, we are adding up all the variability represented in the set of scores, and then dividing that whole into chunks of even size. We can do the same thing with the deviations in our set of scores.

In the formula for the mean, to find the typical amount of deviation (which we can call the **mean deviation**), we would add up the deviations and then divide by the total number of cases.

When we add up the deviations, positive and negative, though, we get 0, or something exceedingly close to it. Try it. Always and every time (except for rounding error), this is the case. Why? Because by adding deviations together, we are just reassembling what we took apart when we found each difference from the mean. In essence, what we've just done is similar to how we found the mean in the first place: we found the mean by folding all that variability into a single ball and dividing it into even pieces. By adding up all the difference between each score and the mean we do in reverse what found the mean in the first place.

This gives us a good way of checking our arithmetic, actually: the sum of the differences between each score and the mean will always give us 0, or very close to 0. If it isn't, something wasn't quite right somewhere.

So, what do we do? Well, the problem appears to be the negative signs. If we can get rid of *them*, we can add up the differences and be closer to an answer, maybe. One option is to take the absolute values of the differences, or simply to ignore the negative signs, when adding up the differences. The resulting sum, divided by the number of scores, is called the

Table 4.2 Years of Service of Sworn Officers in Your Department

Officer	*Years of Service*	*Difference from Mean (11.85)*
Roberts	10.44	−1.413
Van Slyke	22.92	11.067
Lenning	15.36	3.507
Leneker	20.80	8.947
Adkinson	20.07	8.217
Vollum	20.66	8.807
Garland	19.37	7.517
Brooker	21.34	9.487
D'arcangelis	6.75	−5.103
Halpin	4.06	−7.793
Sklar	3.12	−8.733
Sparks	9.22	−2.633
Barile	4.89	−6.963
Hicks	7.57	−4.283
Wright	4.15	−7.703
Brightman	8.29	−3.563
Barlow	5.81	−6.043
Vandewerker	5.13	−6.723
Lyons	3.00	−8.853
Powell	7.67	−4.183
Hayes	2.88	−8.973
Adelman	11.46	−0.393
Jonker	12.34	0.487
Bennett	14.14	2.287
Klemme	21.63	9.777
Collins	13.17	1.317
Fiebig	15.50	3.647
Cannady	13.95	2.097
Smith	10.51	−1.343
Kretser	19.38	7.527

mean absolute deviation, or **MAD.** The MAD turns out to be less useful than we might want, although it does have its uses. Let's think of the MAD as:

$$\text{MAD} = \frac{\sum\left(\left|x - \bar{x}\right|\right)}{n}$$

The MAD is a defensible measure of dispersion, although it lacks the probabilistic qualities we are going to need, as we will soon see. Another way of saying this is that the MAD doesn't allow for comparison of variables to each other as easily as we need.

The Variance and Standard Deviation

Another mechanism for ridding ourselves of those pesky negatives is to square the differences. In general, this is the preferred method. It was preferred by early statisticians, possibly because of the influence of Euclidian geometry on how they thought, but the squaring method gives us measures that are more useful probabilistically than the MAD is. The first of these is called the variance.

$$\sigma^2 = \frac{\sum (x - \bar{x})^2}{n}$$

So, the variance is the average squared mean deviation from the mean. **Each observed score is subtracted from the mean, which we can think of as the score we might expect any given score to be if we didn't actually know it (the mean is the best guess of any given value).** The variance is a good measure of variability, but it is not expressed in the same units as the original data (or it is expressed in units squared, which isn't as intuitive as the units themselves would be). If we take the square root of the variance, however, we get a measure that is in the same terms as the variable.

$$\sigma = \sqrt{\frac{\sum (x - \bar{x})^2}{n}}$$

This measure is the **standard deviation.** It provides for us an expression of variability that is presented in the same measurement units in which the data are presented. Now, *for a sample, though, we use the formula*:

$$\sigma = \sqrt{\frac{\sum (x - \bar{x})^2}{n - 1}}$$

In order to correct for sample degrees of freedom (which I will explain in greater detail later), we subtract one from the sample size.

It bears noting that the standard deviation, either for a population (where instead of an $\bar{x}$ we would use a μ to stand for the population parameter mean) or for a sample as above, the mechanics of the formula carry a mark that will be repeated again and again throughout this book. The difference between x and $\bar{x}$ is in essence the signal that interests us. The sample size, (n or $n-1$), though not purely a measure of "noise," nonetheless provides something of a sense of context or background that is akin to noise. It serves us well to note the anatomical structure here as it will be repeated again and again and again throughout this book and through much of inferential statistics.

The Humble but Mighty Z Score

The standard deviation can do even more for us. Say we want to compare scores in a range to some sense of likelihood, or even to each other in terms of likelihood. No trick. We can use z scores.

The z score is a *fantastically* powerful tool. It allows us to compare each outcome in a set of scores to each other, and even to other scores in other distributions, regardless of scale!

Table 4.3 Data for Our z Score Analysis

Officer	x	$\bar{x}$	$x-\bar{x}$	$(x-\bar{x})^2$	$z=\dfrac{x-\bar{x}}{\sigma}$
A	4				
B	13				
C	5				
D	3				
E	5				
F	3				
G	4				
H	9				
I	4				
J	2				
K	0				
L	1				
M	4				
N	3				
Σ	60				

How is that possible? I'm glad you asked! Take a look at the formula for calculating a *z* score:

$$z=\frac{x-\bar{x}}{\sigma}$$

It is another division formula with subtraction in the numerator – finding the difference between observed (x) and expected ($\bar{x}$) values. I detect a theme! Notice also the denominator is the standard deviation. **Note: whenever we use division like this, we are expressing the numerator in the units of (that is, in denominations of), the denominator. The result of this equation then can be read as the number of standard deviations above (positive) or below (negative) the mean for the score in question. It has been standardized.** In this case, we are expressing the difference between a given score and the mean in units of standard deviations.

If you recall, we looked at histograms earlier in the chapter on displaying data. It is common to report frequencies for a continuous variable using a frequency histogram. You may have already noticed this particular bell-like shape. For any frequency histogram, we can find the mean and the standard deviation, as we have already done. Well, with a *z* score, the shape of the frequency histogram does not change from the histogram of raw scores, but instead of having a mean and standard deviation in terms of the original units (e.g., wombats per square meter, milligrams of mayonnaise on a sample of sandwiches), the mean of a standardized distribution is 0 and its standard deviation is 1. This is because each score has been expressed in terms of the number of standard deviations it is from its mean. That means we can compare any score in the distribution to any other score in the distribution, **or in any other distribution** for that matter. We get to keep the information each point represents, but we get to express it in a way that transcends scale. Now *that* is pretty slick. We will talk a little more about *z* scores later on.

We talked earlier about signal-to-noise ratios. Well, perhaps the first truly evident example of a signal-to-noise ratio is the z score: the difference between each score and the sample mean represents the signal of interest (much like a sample standard deviation is derived), but instead of n (with a population) or $n-1$ (with a sample) for context, we use the standard deviation, the average amount of inherent variance in the sample: noise. A z score, then, expresses the amount of difference from the mean any given score is (how typical or ... potentially interesting ... that score is) in the context of the average amount of variance in that sample. Signal and noise.

Let's try to create some z scores to see how they work.

First, our data. In the first column (x) is a series of numbers. These are our values for x for each of 14 cases. In this instance, let's say these numbers represent the number of field contacts reported by each officer ($n = 14$) on duty during a daytime 12-hour shift. We will treat this as a sample of officers, then.

Looking back at the formula, it is clear that we need a few ingredients for our z scores. First, we need a mean and standard deviation for the range of scores. Then, we need to find the difference between each score and the mean. From there, the rest is easy. First, though, the mean.

$$\bar{x} = \frac{\sum x}{n}$$

Summing the number of field contacts, we get 60 field contacts. There are 14 officers, so the average number of contacts per officer ($\bar{x} = 60/14 = 4.2857$). Incidentally, yes, early in the calculation process, we keep precision; here, we kept precision to four decimals, or the ten-thousandths place. Rounding early in the process increases the blurriness of our values; it is usually best to round at the end, and not so much at the beginning.

Next, we need to derive the values for $x - \bar{x}$. Simply subtract each score, x, from the mean ($\bar{x}$) (see Table 4.4). Once we find this difference for each value, we then square that difference. We are doing the difference and then the squaring separately for a reason.

Now we are simply squaring the difference we found for each case (each row, making "officer" our unit of analysis). See Table 4.5 for the results.

OK, so Table 4.5 presents almost everything we need for a standard deviation. Basically, we have the numerator for the fraction at the heart of the standard deviation formula; we squared each difference, then added them up across all scores. That's what the 138.8571 is. Again, we are keeping precision at this point because there is no reason to lose it yet. Next, though we wind things up, let's look again at the formula:

$$\sigma = \sqrt{\frac{\sum (x - \bar{x})^2}{n - 1}}$$

And let's plug in some numbers and wind it up:

$$\sigma = \sqrt{\frac{138.8571}{14 - 1}}$$

Table 4.4 Contact Data with Mean and *x* Difference from Mean Adde

Officer	*x*	$\overline{x}$	$x-\overline{x}$	$(x-\overline{x})^2$	$z=\frac{x-\overline{x}}{\sigma}$
A	4	4.2857	−0.2857		
B	13	4.2857	8.7143		
C	5	4.2857	0.7143		
D	3	4.2857	−1.2857		
E	5	4.2857	0.7143		
F	3	4.2857	−1.2857		
G	4	4.2857	−0.2857		
H	9	4.2857	4.7143		
I	4	4.2857	−0.2857		
J	2	4.2857	−2.2857		
K	0	4.2857	−4.2857		
L	1	4.2857	−3.2857		
M	4	4.2857	−0.2857		
N	3	4.2857	−1.2857		
Σ	60				

Table 4.5 Squared Deviation

Officer	*x*	$\overline{x}$	$x-\overline{x}$	$(x-\overline{x})^2$	$z=\frac{x-\overline{x}}{\sigma}$
A	4	4.2857	−0.2857	0.08162449	
B	13	4.2857	8.7143	75.9390245	
C	5	4.2857	0.7143	0.51022449	
D	3	4.2857	−1.2857	1.65302449	
E	5	4.2857	0.7143	0.51022449	
F	3	4.2857	−1.2857	1.65302449	
G	4	4.2857	−0.2857	0.08162449	
H	9	4.2857	4.7143	22.2246245	
I	4	4.2857	−0.2857	0.08162449	
J	2	4.2857	−2.2857	5.22442449	
K	0	4.2857	−4.2857	18.3672245	
L	1	4.2857	−3.2857	10.7958245	
M	4	4.2857	−0.2857	0.08162449	
N	3	4.2857	−1.2857	1.65302449	
Σ	**60**		**0.0002**[a]	**138.857143**	

[a] As I noted, the sum of the differences between *x* and $\overline{x}$ will sum to 0. Well, this number isn't exactly 0, small though it may be. The reason it is not exactly 0.0 is because of rounding "error."

$$\sigma = \sqrt{\frac{138.8571}{13}}$$

$$\sigma = \sqrt{10.6813}$$

$$\sigma = 3.268.$$

Table 4.6 Making z Scores

Officer	x	$\bar{x}$	$x-\bar{x}$	$(x-\bar{x})^2$	$z=\dfrac{x-\bar{x}}{\sigma}$
A	4	4.2857	−0.2857	0.0816244	−0.08742
B	13	4.2857	8.7143	75.9390245	2.66637
C	5	4.2857	0.7143	0.5102245	0.21856
D	3	4.2857	−1.2857	1.6530245	−0.39339
E	5	4.2857	0.7143	0.5102245	0.21856
F	3	4.2857	−1.2857	1.6530245	−0.39339
G	4	4.2857	−0.2857	0.0816245	−0.08742
H	9	4.2857	4.7143	22.2246245	1.44246
I	4	4.2857	−0.2857	0.08162445	−0.08742
J	2	4.2857	−2.2857	5.2244245	−0.69937
K	0	4.2857	−4.2857	18.3672245	−1.31132
L	1	4.2857	−3.2857	10.7958245	−1.00535
M	4	4.2857	−0.2857	0.0816245	−0.08742
N	3	4.2857	−1.2857	1.6530245	−0.39339
Σ	60		0.0002[a]	138.857143	

[a] As I noted, the sum of the differences between x and x will sum to 0. Well, this number isn't exactly 0, small though it may be. The reason it is not exactly 0.0 is because of rounding "error."

And yes, we rounded here at the very end. So, our standard deviation is 3.268. We are all set to make some z scores! Recall that the formula for a z score uses $x-\bar{x}$, **not** $(x-\bar{x})^2$. We don't want to be rid of the positives and negatives for our z scores. All we need to do now is to divide each of the numbers in the $x-\bar{x}$ column by the sample standard deviation (Table 4.6).

In this last column, then, are our z scores. They express the difference from the mean for each individual score in terms of numbers of standard deviations. This is *wildly* useful. Some are negative, meaning that they lie a certain number of standard deviations below the mean, whereas some others are positive, indicating that they are situated above the mean. Now here's a question: do any of these z scores stand out to you? How so?

Outliers...?

There is a problem, though. Standard deviations are important because they give us the chance to calculate the highly versatile z score, as we have just seen. But like the mean, a standard deviation is subject to influence by extreme scores, or even **outliers**, scores that fall well outside the typical range (we will discuss how to identify outliers in a moment). This makes sense because the mean, itself potentially subject to "volatility" in low-n situations, is a key ingredient in a standard deviation. As so many of us do, the mean passes on its insecurities to its child.

But what if we had a different measure of centrality, one that isn't yanked in directions by outliers? Recall from Chapter 3: we actually do. The median may not make use of every bit of information of a variable, but it hangs tough against the influence of outliers. So, what if we decided to put this to work for us? This is just what Leyes and colleagues (2013) suggest we do in certain cases. When n is small (meaning there are only a few cases), and there is an extreme outlier, they recommend we find the *median* absolute deviation from the *median*. First, we would find the median for our scores, then we would find the absolute value of the deviations from the median:

$$|x - \overline{\mathrm{Md}}|$$

Finally, we would identify the median of the deviation scores. Consider the example offered by Leyes and colleagues (2013). They offer a made-up set of $n = 8$ scores: 1, 3, 3, 6, 8, 10, 10 and 1,000. The mean of the scores is 130.13, and the standard deviation is 328.8. One way of identifying outliers is through the use of standard deviations: we could say that any score that is more than three standard deviations above or below the mean is a *candidate* to be an outlier. In their example, although the 1,000 is clearly a strangely high score, three standard deviations ($328.8 \times 3 = 986.4$) above the mean (130.13) is 1,116.53. *That means the 1,000 outcome would not be an outlier* if we used this method.

Hmm. Well, the median absolute deviation from the median gives us a way to justify calling Leyes' weird score an outlier. And doing so would be of use because it does seem to be having an outsized impact on our ability to talk about typicality and dispersion in any clear way. Notably, that 1,000 outcome might be real and meaningful, and we should not want to excise it merely because it is weird. We need to have a good reason before we contemplate lopping off scores.

The median position is between the 4th and 5th scores (6 and 8). We average them and get 7 for the median. Subtracting the median from each score, we find the median deviation to be 3.5. If we then divide each deviation from the median, we find that for the 1,000, the median absolute deviation from the median is well above the 3 deviations mark (it is more like 283.7 deviations), while the rest of the scores present us with much more sedate differences from the median. If we have the luxury of more scores, though, the benefits of the median absolute deviation tend to be fewer.

The usual method of using standard deviations (or z scores) ably identifies the high score as a potential outlier in our sample data. In fact, using the median absolute deviation method, we might identify not one, but *three* potential outliers, two of which do not seem so terribly extreme.

Coefficient of Variation

$$\mathrm{CV} = \frac{\sigma}{\bar{x}}$$

The coefficient of variation (CV) is another measure of variability, but one that operates relative to the sample mean. It is a ratio of the standard deviation to the mean, interpretable as a percent (if we multiply the result by 100 – note, though, that the CV is not bounded on the high end at 100, so it isn't a proper percent [see, e.g., Bedeian & Mossholder, 2000]). For ratio-level variables, it is perhaps just slightly slicker than a standard deviation because the result is not expressed in units of the variable, so it is more comparable across variables with different units. It is not exceedingly common in social science for the simple reason that continuous variables that are truly ratio are relatively uncommon; CV should not be used with data that are not properly ratio. Smaller CV values indicate a more consistent sample, meaning that the variation is relatively snugly set around the mean. Context for analyses definitely matters here (see, e.g., Bedeian & Mossholder, 2000), but a CV = 0 indicates total uniformity (all scores are the mean). Ideal CV scores are nonzero (because a CV = 0 indicates the variable is a constant) but CV values below 10, or above 30, might indicate some problem exists with the data (again, though, these are not bright-line standards, particularly in the social sciences).

Skew

As we said before, not always are our distributions symmetrical, so it helps us to have a measure of wonkiness of a distribution. A normal distribution is even on both sides, piled up in the middle. Such a distribution is not skewed. When a distribution has to lean to one side or another, we say it exhibits skew. Typically, we describe skew according to the value of skewness scores – as positive or negative (see Figure 4.1).

Now, like many statistical concepts, there are a few ways to understand (and thus to calculate) skew. Karl Pearson crafted two such ways. It seems he was somewhat fonder of his first coefficient of skew, which is perhaps the simplest of all to envision:

$$Sk_1 = \frac{\bar{x} - \text{Mode}}{\sigma}$$

... or simply the difference between the mean and the mode divided by the sample standard deviation. Notice: signal and noise again. We can see rather clearly how this works:

What Karl Pearson has done here is rather clever. We know from Chapter 3 that means are more susceptible to being tugged in different directions by relatively extreme scores, whereas the mode and the median are more resistant to that same tug. If we want to know something about the shape (i.e., lean) of a distribution, it makes sense to leverage the relatively peripatetic nature of the mean relative to other measures of centrality; why not put those traits to good use?

Something else we can see in Figure 4.2 is (one reason) why the skew shapes have the names (i.e., positive and negative) they do. The positive skew shape, with the tail pointed off to the right, is positive because a bigger number (the mean) is subtracted from a smaller number (the mode), leaving a positive number after the division. The opposite is true for a negative skew: a smaller number (the mean) is subtracted from a larger number (the mode) for the numerator. For a normal distribution, the skew is equal to or at least very near zero.

One of the issues with the Pearson's first coefficient is that if the mode isn't strong, meaning that there are only a few instances of the modal value (recall that the mode is just the

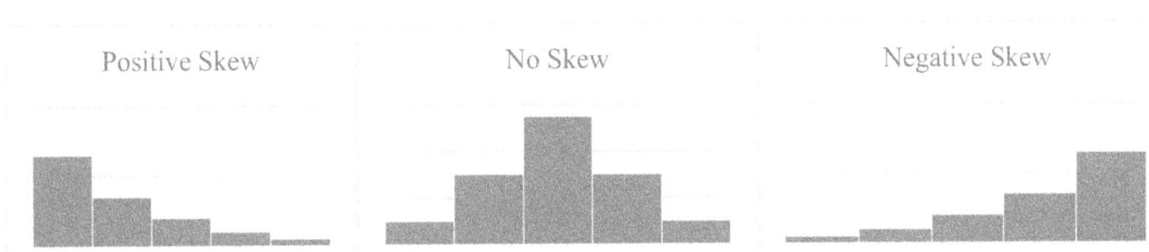

Figure 4.1 Skewness.

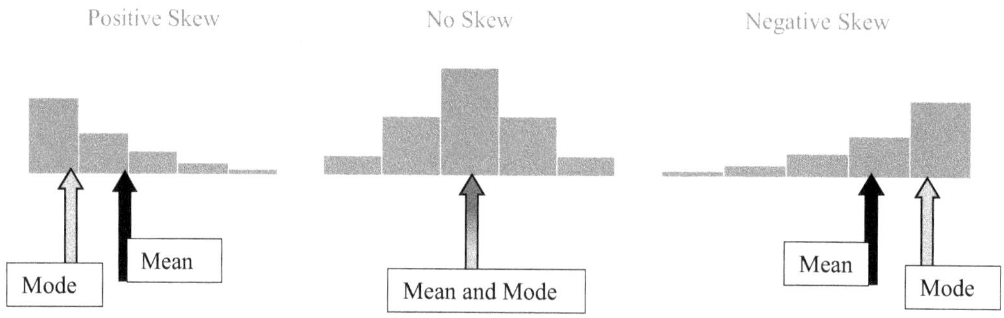

Figure 4.2 Means and Modes in Different Skew Environments.

most frequent score), then the measure is less precise. One way to compensate for a weak mode is to use Pearson's second coefficient of skew.

Pearson's (second) coefficient of skew is:

$$Sk_2 = \frac{3(\bar{x} - \text{Md})}{\sigma}$$

So instead of using the mode, we use the median (and multiply the difference by 3). That is to say, we take the sample mean and subtract the median, multiply the difference by 3, and then divide that product by the sample standard deviation. Notice again: signal and noise, just like before.

Typically, though, many statistical software packages default to a different measure of skew, usually called the Fisher–Pearson coefficient of skewness (e.g., Doane & Seward, 2011). Although it and its variants are more common, it may not be as satisfyingly intuitive:

$$g = \frac{\frac{1}{n}\Sigma(x - \bar{x})^3}{\left[\frac{1}{n}\Sigma(x - \bar{x})^2\right]^{3/2}}$$

A similar, somewhat less intimidating scale for skewness is given by Warner (2008):

$$\text{Skewness} = \frac{\Sigma(x - \bar{x})^3}{(n-1)\sigma^3}$$

where

x is each score in the sample

$\bar{x}$ is the sample mean

n is the sample size

σ is the sample standard deviation.

Scores for the Fisher–Pearson skewness coefficient and the skewness scale mentioned by Warner above are similar to the Pearson measures above in that a perfectly symmetrical distribution will have a skewness coefficient of 0, and positive and negative scores indicate wonkiness in the same fashion as Pearson's measures. The *cubed* sum of each $x - \bar{x}$ pair *brings back the negatives* we lost when we squared this comparison when calculating standard deviation. In this case, the positives and negatives actually work to our advantage by giving us a sense of distributional balance. Again, the numerator contains the signal, and the denominator represents the measure of noise (or of freedom for variance).

Kurtosis

And it should be no surprise that we find much the same signal-to-noise dynamic in our measure of kurtosis. Kurtosis is the degree of probability density of the shoulders and tails of the distribution. A platykurtic distribution is, in essence, all shoulder and no tail. A normal distribution, like the z distribution, has both some shoulders and some tails, and is described as mesokurtic. A leptokurtic distribution has narrow shoulders offset by thicker tails.

Some treatments of kurtosis emphasize the pointiness of distributions alone as being the focal concern of kurtosis. Such treatments are less precise. Kurtosis has less to do with the "peakiness" of the distribution itself than it does about the relative probability densities of the shoulders as compared with that of the tails of the distribution. Extreme scores are more prevalent, for example, with leptokurtic distributions and have an outsized influence on the kurtosis score.

It is fair to think of kurtosis, then, as being in part a measure of the consensus around the mean. If the mean were an object in space, kurtosis might be analogous to a measure of the depth of the gravity well created by that object (Figure 4.3).

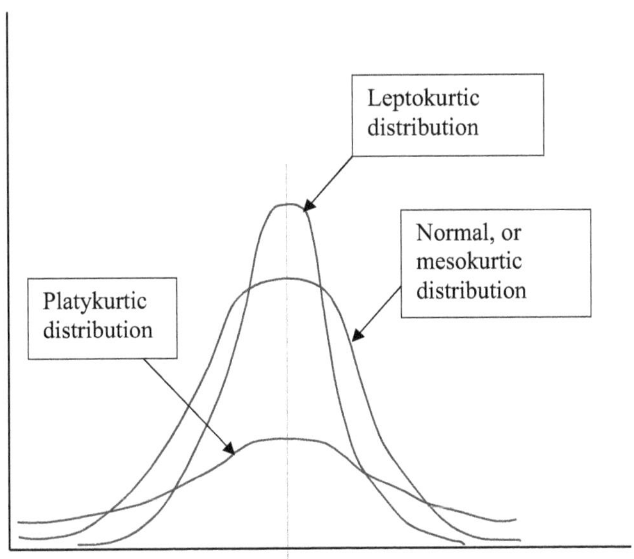

Figure 4.3 Kurtosis.

Kurtosis is measured (Warner, 2008):

$$\text{Kurtosis} = \frac{\Sigma(x-\bar{x})^4}{(n-1)\sigma^4}$$

where

x is each score in the sample
$\bar{x}$ is the sample mean
n is the sample size
σ^4 is the sample standard deviation to the fourth power.

A normal distribution that approximates a z distribution would have a kurtosis of 3.0. Alternatively, we might use (Warner, 2008):

$$\text{Excess kurtosis} = \frac{\Sigma(x-\bar{x})^4}{(n-1)\sigma^4} - 3$$

to describe the degree to which a distribution differed from a normal distribution. By subtracting 3 as we do in the equation above, a normal distribution would have a score of 0; any deviation in either direction (a negative score indicates platykurtosis, and thus thinner tails; a positive score would indicate thicker tails and narrower shoulders and thus be leptokurtic. We call this *excess kurtosis* (e.g., Warner, 2008: 149).

What Do We Do about Skewed Or Kurtotic Data?

In cases where there is a need to correct for an amount of skew, we can perform different **transformations** on the skewed variables. These transformations almost always make interpretation of results more involved and less concise, but transformations can serve our ends under certain circumstances (Table 4.7).

In essence, what this means is that instead of examining a distribution of variable x, which, say, has a strong positive skew, we might look for normality (bell-shapedness) in a distribution of the Log10 of x. And of course, the Log10 of x is not nearly as clear as x itself – after all, what would such a thing actually mean? Interpretation of transformed variables becomes rather dicey, but if normality is important (and there are times where it is), transformations can save our proverbial bacon.

Conversations about transformations are for a more advanced treatment on statistics. For the moment, though, it is sufficient to note merely that there are ways out of situations where data do not adhere to assumptions about univariate (or bivariate) normality.

Table 4.7 Skewness and Transformations

If your variable is...	*...try transforming by calculating:*
Somewhat positively skewed	the square root of x
Very positively skewed	cube root of x; the Log10 of x
Somewhat negatively skewed	$k - x$; the square root of $(k - x)$
Very negatively skewed	cube root of $(k - x)$; the Log10 of $(k - x)$

k = a constant from which each score is subtracted such that the lowest score in the range is equal to 1.0.

Summary

Variety, they say, is the spice of life, and this old yarn is no truer anywhere else than it is with regard to statistics. But we must be clear-eyed about the nature of the variability before us if we plan to make magic with our statistics. In this chapter, we have examined several cornerstone concepts related to the description of variation of a variable. The variance and the standard deviation will be used frequently as a crucial ingredient in much of what is to come. Variation ratios and coefficients of variation are useful devices but are relatively less common in human sciences and are not relied upon as an ingredient of most other statistical techniques. Skew and kurtosis are mightily useful descriptive terms for the understanding and proper utilization of variables' variability as they present themselves to us. Strongly skewed or ragingly leptokurtic variables can give us some fits, but almost always there is some adjustment that can be made to make a wonky variable defensibly less so.

Main Take-Aways

- Understanding variability within our data, be our data a sample or a complete population, is as vital as understanding the nature of centrality or typicality. Indeed, these two understandings work hand-in-glove.
- Multiple measures of dispersion exist, from the range and variation ratio to standard deviation, *z* score, skewness and kurtosis.
- Levels of measure play a crucial role in determining which expression of dispersion is most meaningful.
- Shortcomings of some statistical measures can be turned into capacities.

Key Terms (in order of appearance)

range
variation ratio
mean absolute deviation
variance
standard deviation
z score
outliers
coefficient of variation
skew
kurtosis
transformations

Chapter 5

Basic Probability and the Normal Curve

Before we get too far down the road, we need to make a slight detour to talk about the basic rules of probability. These rules are helpful in developing an understanding of the whys behind the reasoning we use to draw inferences about **populations** from **samples**. This will not be a full-on treatment of all the rules of probability. Instead, it will approach probability from a need-to-know perspective: the stuff we need to know to understand how to operate the statistical tools we will talk about later on.

More specifically, we want to be able to reason our way to conclude things about a population based on the relatively limited information available from a sample from that population.

Remember in Chapter 1 how we talked about signal and noise? Information that stands out from that noise might well be signal, but to be able to separate signal and noise, we need some reference point for our noise. Basic probability gives us a framework for defining that noise.

You see, there is a natural variability to things in our world. Sure, people and objects obey the laws of physics, but outcomes that might interest us, though they also obey physical laws, are unpredictable to a degree. We know this just from living in the world. Well, this variability, this rattle-and-hum of things in the human universe, should be measured to the degree possible because it has the potential to obscure understandings about outcomes and their causes. Probability theory is one way of doing just this. Let's say we were concerned with property crime in a community. Taking matters seriously, the local police undertake several programmatic efforts to intervene against business break-ins and motor vehicle thefts. After a period of time, they note a 4% drop in targeted property crime. Time to celebrate, right? Well, let's wait on ordering the cake, because we know that property crime (and a host of other things in the human universe) are not static. They vary over time, often within a measurable range, and without evident cause. Looking more closely, we might find that the month-to-month change in property crime is on the order of an increase or decrease of as much as 6%. A 4% drop, then, might just be noise. It would be impossible to be certain, of course, but if the change before us – that 4% about which we were about to celebrate – is within the bounds of natural variation, we would be inviting calamity to assume that change was the result of something we did.

Have you ever bought a lottery ticket? You might know the chances of winning are exceedingly small, possibly on the order of 1 in 292 million (which is about the odds of winning the Powerball lottery at the time of writing). To talk about the probability of something in this way is to make a statement about the relative likelihood of an event happening *under conditions of randomness*. Randomness is not willy-nilly-ness. It is, well,

DOI: 10.4324/9781032638591-5

systematically non-systematic...ness. The simplest most precise way to think about randomness is that *each possible outcome has an equal likelihood of occurring.* If we throw a fair (meaning it isn't weighted) six-sided die, each side of the die has an equal chance as every other of landing up. For a fair coin, each side of a tossed coin has an equal likelihood of landing up, too. If we have a population of people and we want to draw a random sample from it, each member of that population must have an equal (even if vanishingly small) likelihood of being selected. In research parlance, we talk about sampling or group assignment that uses a method that assures equal probability of selection or of assignment as "EPSEM" (**E**qual **P**robability of **SE**lection **M**ethod).

In probability terms, each time we gather a set of observations about something in the world, like posing a set of questions to someone and measuring her responses or observing the outcomes of five throws of a six-sided die, that is an **experiment.** Each experiment has a number of potential outcomes, sometimes called **degrees.** The particular outcome or outcomes (say, one of the five throws of that die) that interest us is called an **event.** The full list (or set) of those potential outcomes or events is called the **sample space.** The sample space for a single coin flip has two members: heads and tails. The sample space for a single throw of a fair die is 1, 2, 3, 4, 5 and 6, and the sample space for, say, an interview question that uses Likert-type items (i.e., strongly agree, agree) would be the response options provided for each question.

We can quantify the likelihood of any given event using the following bit of logic:

$$\text{Probability of an event} = \frac{\text{times the outcome (event) can occur in an experiment}}{\text{total number of possible outcomes in the experiment}}$$

Say we roll one fair die, and we want to get a 6. There is one side of the die with six dots, and there are a total of six sides on the die. The probability of getting a 6 on a single throw of a die is:

$$\text{Probability of a 6 on one throw of a 6-sided die} = \frac{1}{6}$$

We can also express this number as a decimal, which is a little more useful:

$$\text{Probability of a 6 on one throw of a 6-sided die} = \frac{1}{6} = 0.1\overline{66}$$

as 1 divided by 6 gives us 0.166666. This number is the **probability** of our event, which gives us some sense of its **relative frequency.** The probability of an event is often denoted with a p. The probability of a 6 on a single throw of a fair die is 0.16667. We say the same thing when we point out that we expect to get our 6 (or any other desired outcome on that die) about 16.67% of the time *if all things are fair.* A fair die would have the probability of every possible degree or member of the sample space to be the same; if the probability of a 6 is 0.16667, then on a fair die, the probability of a 2 or a 5 or a 1 would each be 0.16667. This "if" is huge, of course, and relates to the whole logic of how we use statistics to examine things in the criminological world. More about this in Chapter 6.

Probability, as you can see, is expressed as a decimal. That means it can fall between 0 (the event will never occur) and 1 (the event is a lock, a certainty). We know that the probability of a 6 is 0.16667. The probability of *not* getting a 6, called q, is the **complement** of the probability of the event (or $1 - p$). That means the probability of rolling something other than a 6 is $1 - 0.16667$, or 0.8333. We could also calculate the probability of getting a 1, 2, 3, 4 or 5 directly, which would be 5/6, or 0.8333.

Well, we can take our coins and our dice and throw them multiple times, as we might do in craps or Yahtzee. And if we happened to have a little cash riding on the throw (not that I condone such a thing), we might want to have a more precise sense of its relative likelihood. We might also combine outcomes for our events. For example, we might be interested in getting an even number roll, or perhaps a 4 one die and then a 3 on a different die. There are some concepts and rules that help us reason our way to deriving probabilities for these situations.

Conditional Probability

There are times where the events that interest us are interdependent. Things in the world are like that. *We* are like that. When events are linked in some way, as when the probability of one experiment is shaped by the outcomes of another experiment, we are thinking in terms of conditional probability.

Replacement and Non-Replacement

The most basic version of a dependent event is what we would refer to as **non-replacement.** Say we have a sack with 12 marbles in it. Five are red, seven are blue. The probability of reaching in blindly and drawing a blue marble is $\frac{7}{12} = 0.58333$. Of course, that also means that the probability of drawing a red marble is $1 - 0.58333$, or 0.41667. Say we do precisely that: one of us reaches into the bag and yanks out a blue marble, places it in her slingshot and fires at a suspicious-looking Pepsi bottle.

Gone. Nice shot. Now we have *11* marbles in the sack: five red and *six* blue. The probability of drawing a blue marble now is different from it was before, because the proportion of blue marbles to the total is different. There are now both fewer blue marbles and fewer marbles overall. But the probability of drawing a *red* marble is *also* different, even though the number of red marbles has not changed. The probability of grabbing a blue marble is now $\frac{6}{11} = 0.54545$, making the probability of grabbing a red one $1 - 0.54545$, or 0.45455. That's something like an 8.3% increase in the probability of getting a red marble $(1 - (0.41667/0.45455))$, achieved only by grabbing a blue one.

If, instead, our friend had not rocketed our marble at the crockery without asking first, we could have put the marble *back* into the sack, and the next draw would have the same probability profile as the first one.

Addition Rule

Say we want to know the probability of getting a 1 or a 2 on a single throw of a die. Now you can't get both a 1 and a 2, so we would say these outcomes are **mutually exclusive:** either we get a 1 or a 2 or we do not. The probability of getting a 1 is 0.16667, just like the probability of getting a 6. The same is true for getting a 2. The addition rule tells us that if we have mutually exclusive outcomes, the probability of getting one **or** another of them is determined through addition.

The probability of a 1 is 0.16667 (or thereabouts, due to rounding). The probability of a 2 is the same: 0.16667. The probability of rolling *either* a 1 *or* a 2 is:

$$p(1 \text{ or } 2) = 0.16667 + 0.16667 = 0.3334$$

Of course, we could also go about this by saying the probability of getting a 1 is 1/6, as is the probability of getting a 2. 1/6 + 1/6 is 2/6. 2/6 simplifies to 1/3, or 0.333....

Multiplication Rule

Now, if the event that interests us is a more particular outcome: say instead that we wanted to know the probability of getting a 1 and then a second 1 on a second roll. We can calculate that probability by multiplying the probability of the first event by the probability of the second event. As the product of a decimal times a decimal is going to be a smaller decimal, this only makes sense: we are talking about a less likely event.

The probability of a 1 is 0.16667. To get a 1 *and then another* 1 on a second roll, the probability is:

$$p(1 \text{ and } 2) = 0.16667 \times 0.16667 = 0.02778$$

Such a thing as two consecutive 1s can be expected to occur due to chance alone only 2.778% of the time. That's pretty unlikely. The probability of two consecutive 1s *not* happening is its complement, or $1 - 0.027778 = 0.9722$.

Toward a New Capacity

All of that is fancy reasoning, to be sure, but the things that interest us from a criminological perspective are far too complex to be touched by simple conditional probability. The good news, though, is that this reasoning can be ramped up into a grander system. This grander system is called the binomial distribution, and it is the basis for much of Ordinary Least Squares statistical reasoning used in this book from here forward.

The Binomial Distribution and the Normal Curve

So now we have set up probability as a way of measuring noise or expected natural variability. Let's turn our attention to a slightly more useful question that we will address using the change in our pocket. Well, we just need one coin.

We are going to flip that coin five times. We want to be able to measure the probability of all of the possible outcomes that could occur: we could get no heads, one head, two heads, three heads, four heads or five heads. Our rules discussed above give us some help in figuring out how to determine the answers, but the actual formula is somewhat more complicated than either rule. We can figure out the likelihood of each of the possible events (outcomes) for an experiment using the binomial probability formula:

$$P(x) = \frac{n!}{r!(n-r)!} p^r q^{n-r}$$

where

n = the number of trials in the experiment (in this case, the number of coin flips)

r = the number of outcomes that interest us, which is often called the number of successes

p = the probability of success

q = the probability of failure (or not success)

! = factorial (a number times the number below it, times the number below *it*... $5! = 5 \times 4 \times 3 \times 2 \times 1$).

Translated into English, this formula says that the probability of an outcome equals the factorial of the number of trials divided by the factorial of the number of successes times the factorial of the number of failures (that's the *n-r* bit), times the probability of success to the power of the number of successes, times the probability of failure to the power of the number of failures. Yuck.

OK. It is really quite a bit less complicated than it might appear. Think of this formula as operating in two parts. The first part, the fraction with the factorials, works to count the number of ways our event, our outcome of interest, can happen. The second part, the p and q part in the back, expresses the probability that each of these outcomes might occur. We simply multiply the number of ways it can happen by the probability of each happening, and poof!

Let's take our example of five heads. What is the probability of getting exactly two heads when we flip a coin five times? The first thing we do is pop in the values.

$$P(x) = \frac{5!}{2!(5-2)!} 0.5^2 \times 0.5^{5-2}$$

For the first part, we convert the factorials into whole numbers; for the second part, in the case of a fair coin, the probability of success (heads) is 0.5, as is the probability of failure (tails). This second term operates like a weight – it won't change even though the exponents do. It can be understood as the independent probability of deriving one success given the number of trials. In this case, that second term is the probability of getting any particular outcome over the course of five trials.

$$P(x) = \frac{120}{2\ (3)!} 0.5^2 \times 0.5^3$$

Remember that the two terms, the p (0.5^2) and the q (0.5^3) written as they are above, simply means that we should multiply them:

$$P(x) = \frac{120}{2 \times 6} \times (0.25 \times 0.125)$$

We can get a bit closer by cleaning up the denominator a little and by doing the multiplication in the probability bit:

$$P(x) = \frac{120}{12} \times (0.03125)$$

Then we turn the improper fraction into a whole number:

$$P(x) = 10 \times (0.03125)$$

At this point, it is all over but the shouting:

$$P(x)=0.\ 3125$$

So, of the total number of possible outcomes on a fair coin flipped five times, exactly two heads can happen in ten ways. Each two-head outcome has the probability of 0.03125. Having ten events, with each of them having a probability of 0.03125 means that exactly two heads will occur only about 31% of the time when we flip a coin five times. If we do the same thing for each of the possible outcomes, we get a table that looks something like this:

One can't help but notice the symmetry here, right? Well, that symmetry comes from the two moving parts, one in each portion of the equation, but in fact the differences in probability *in this case, in an instance where success and failure are equally likely*, result only from changes in the count, as you can see in Table 5.1. *Although the exponents change as r changes, the product in this case doesn't change.* In the first part, the counting part, when *r* approaches the middle of the range of possible outcomes, the number of discrete possible outcomes increases and the fraction becomes, well, "more improper" (the denominator gets smaller), and that is what ends up determining the probability for each outcome. This just makes sense: getting only one head or getting four heads are both unlikely events; getting 2 or 3 heads can happen in a lot of ways – ten to be exact.

We can think of our formula as being a more complicated application of the addition rule of probability. Why, you may ask, is it an application of the addition rule when it is clear that we're using multiplication? A wonderful question. The answer is that we are multiplying a probability (that's the second chunk of the equation) by the number of times success can arise, which is a way of adding. If we were multiplying a probability by a probability, you'd be absolutely correct. In this case, though, as each set of possible outcomes, say, Heads, Tails, Tails, Heads, Heads, is the same, we multiply the probability of each one times the number of ways it might arise. That is the same thing as adding up the probabilities of each of the times when two heads can happen on five flips. As there are ten sets of heads/tails outcomes on five flips where three heads can happen, though, we just multiply the probability of any one of the five-outcome sets times the number of possibilities.

As we already know, a coin has two discrete possible outcomes, so it is an example of what we call a **Bernoulli experiment**. What happens if we do more flips of that same fair coin ... say ten flips? (Table 5.2).

And again with the symmetry. Peep this: we can represent the same information as a line graph: Boom. That's your bell curve right there, sort of. Alrighty then, let's actually give it a go, shall we? Take that coin in your hand and flip it ten times and count the number of heads you get. Then ask a friend to do the same thing. *Then* track down a stranger or

Table 5.1 Probability Mass Function for Discrete Bernoulli Outcomes for a Fair Coin, Five Flips

Outcome	*Count*	*Probability*	*P(x)*
0	1	0.03125	0.03125
1	5	0.03125	0.15625
2	10	0.03125	0.3125
3	10	0.03125	0.3125
4	5	0.03125	0.15625
5	1	0.03125	0.03125

Table 5.2 Probability Mass Function for Discrete Bernoulli Outcomes for a Fair Coin, Ten Flips

Outcome	*Count*	*Probability*	*P(x)*
0	1	0.000977	0.000977
1	10	0.000977	0.009766
2	45	0.000977	0.043945
3	120	0.000977	0.117188
4	210	0.000977	0.205078
5	252	0.000977	0.246094
6	210	0.000977	0.205078
7	120	0.000977	0.117188
8	45	0.000977	0.043945
9	10	0.000977	0.009766
10	1	0.000977	0.000977

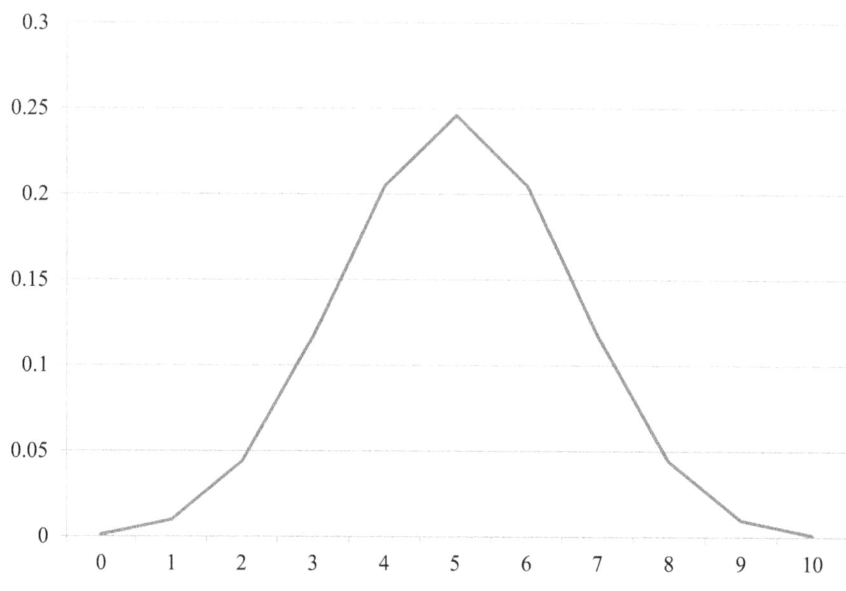

Figure 5.1 Probability Curve, Fair Coin.

someone else, and then someone else. Get as many suckers to help you as you can and keep a tally of how many heads each sucker gets. Make a frequency table of the scores. Then see if you can sketch a bar chart for your outcomes. Does it look like the curve in Figure *x*? How close is it? Do the differences you see matter? That's the natural next question, isn't it?

When we compare our probability curve in Figure 5.1 with the outcomes you and your pals got, we have an essential mechanic in statistical reasoning. The binomial distribution we just made using the probability formula represents an **expected** set of outcomes. The number of heads each of your friends and enemies derived was our **observed** or empirical outcome.

Think about it: if you have an observed outcome of some kind, like the number of heads you got on ten flips, the next logical question is, "How meaningful is this?" Meaningfulness is a thing we might grasp more fully if we shared an afternoon with Viktor Frankl, Simone Weil, Johann Metz and Paul Tillich, but sadly such a moment is not possible. What they might tell us would be of far greater import, perhaps, but for the particular questions to which we turn here, we have statistics. As statistics does not deal at all with meaningfulness (you and I have to bring the meaning to the table), the best we can do is to speak to relative likelihood, to invoke some kind of *comparison*. But again, to do *that*, we need some kind of *context*, some benchmark against which to judge the outcomes we have. The binomial distribution we see here serves defensibly as a set of expected results against which we can compare our observed results – it gives us a brand of context we crave for understanding the results in our hands.

Now all this symmetry bit applies to a fair coin, or a Bernoulli experiment where the probabilities of success or failure are the same. Let's say instead of a fair coin, we have a situation where the probability of success is 0.4, not 0.5. Maybe instead of a coin we are talking about a machine making free throws from the foul line. All that pretty symmetry goes away. Depending on the skill of the machine, the shape of the resulting probability curve changes (see Figure 5.2).

Starting with even odds, like for our coin, as the probability of success decreases, the probability curve shifts increasingly toward the right, evidencing an increasingly positive skew (Figure 5.2). This makes perfect sense if you think about it. The probability of getting a pawltry handful of successes increases as each success becomes less likely. Achieving more successes decreases in probability because success in each trial is harder to come by. The probability of multiple but unlikely successes does not ever become zero, though. This attribute is something I will discuss in more detail later.

Why does that matter for us? For a couple of reasons, actually. First, the fair coin curve we made closely resembles a thing we call the Gaussian, or normal curve. Often it is also

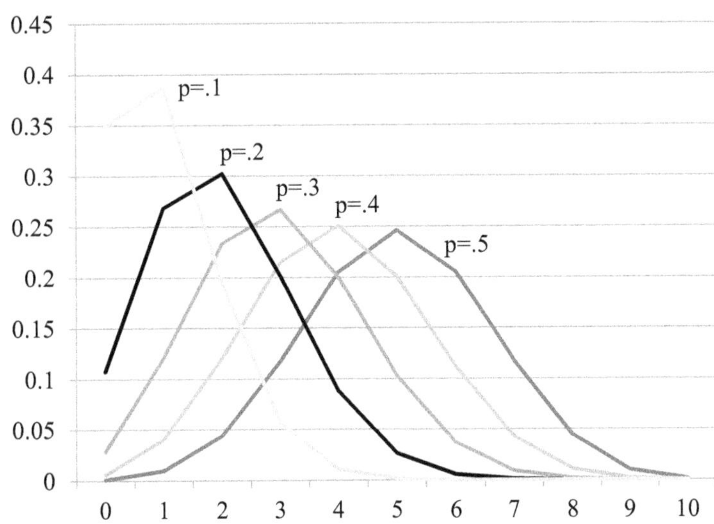

Figure 5.2 Probability Mass Function for Discrete Bernoulli Outcomes, Varying *p*.

called the bell curve because of its bell-like shape. Recall that we talked about the bell curve in Chapter 2. This bell curve is exceptionally useful for helping us approximate natural variation, the rattle-and-hum of things in the world. *We can expect more than a few phenomena in the world to vary on their own, without some outside force acting on them, in a way much like the bell curve approximates.* The bell curve, as I indicated above, gives us something of a measure of noise under normal circumstances. Well, not always are outcomes as symmetrical as a set of fair coin flips, and this leads us to my other key point here. When phenomena in the world vary according to conditions, say, the probability of the event that interests us (like scoring a basket with a free-throw machine), the shape of the probability curve that describes its rattle-and-hum changes. This second point will become increasingly important later on.

Let's look a little more closely at the normal curve. It is here where a number of things we have discussed already, like z scores and the fairness of outcomes, will come together.

Recall from before that a z score represents the number of standard deviations from the mean in either direction a score sits. That's all well, and it seems as if it might be close to something really useful, but again, so what? Well, what if we could actually pin a probability to an outcome based on the number of standard deviations away from the mean it is?

Let's get this out of the way. The normal distribution is derived using this formula:

$$F(x) = \frac{1}{\sigma\sqrt{2\pi}} e^{-(x-\mu)^2} /2\pi^2$$

The good news is that we will not have to operate this formula in this course. It has already been done for us, so we need not revisit it. It is here, though, to make a point: the normal curve represented by this equation is a **probability density function** that approximates the binomial distribution. *It is not the binomial distribution, and it is not an empirical distribution. But it does give us a basis for figuring out an approximate probability of an outcome.*

As I just pointed out, this curve is a probability density function (see Figure 5.3). At the middle is the mean. That midpoint represents the point of even odds: 50% of the curve is below it, and likewise, 50% is above it. Our fair coin balances on this point. Notice the horizontal axis. It is marked in terms of the number of standard deviations away from the

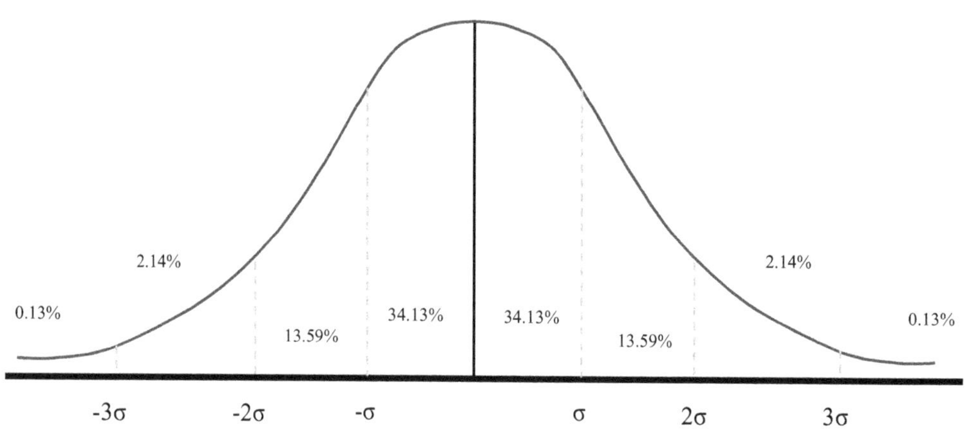

Figure 5.3 Area under the Normal Curve.

mean where each vertical line lands. These vertical lines chunk the curve into sections. It turns out that the area under the curve in each chunk corresponds to a set probability. A score that is exactly one standard deviation above the mean marks a place where 34.13% of the area under the curve exists between the mean and that score. Some normal curve tables indicate the probability value, the percentage of space under the curve, between the mean and that point. The table in the back of the book not only shows the area between the mean and a given score, but it also shows the area beyond z. Depending on our question, either number can be useful for figuring out how likely a score is. We will develop this capacity soon.

So, say we get an empirical finding that interests us, and we want to know how meaningful it is. Again, as we can't necessarily speak to *substantive* meaningfulness, *the best statistics can do is to approximate how likely it is to happen under normal circumstances* (meaning when other things – like the things that interest us – are not acting on it). We need context to answer the question, and we derive that context by determining the probability of the outcome we see (our **observed** outcome). When we standardize our scores, we set up a comparison with a probability density function, so that we can draw some conclusion about likelihood. Let's take an example.

Here is a set of scores and their corresponding z scores. The mean of the set is 6.26667, and the standard deviation is 4.3337 (Table 5.3).

There is a weird score among them, isn't there? That 21 stands out. It is almost 3.4 standard deviations above the mean, and that's pretty far, but the question is *how likely* is that score to occur under normal circumstances? Another way of saying the same thing is to ask how likely it is that our 21 is a member of the same population as all the rest of the scores, given that population has a mean of 6.26667 and a standard deviation of 4.3337?

Because z scores are standard, meaning that regardless of scale or variability, *a given z score in any distribution is about as likely as it is in any other distribution.* Cool, no? That means we can take our z scores for this distribution and hold them up to the probability density function we call the normal curve, and from that derive an estimate of relative likelihood.

Table 5.3 Raw Scores and Corresponding z Scores

Raw Score	*z Score*
5	−0.29228
4	−0.52303
3	−0.75378
5	−0.29228
4	−0.52303
6	−0.06153
6	−0.06153
4	−0.52303
7	0.169216
3	−0.75378
21	3.399713
7	0.169216
6	−0.06153
5	−0.29228
8	0.399966

Table 5.4 Raw Scores, z Scores, Percentile and AUC

Raw Score	*z Score*	*Percentile (Using Excel)*	*Approximate Area from Mean to z (% AUC)*
5	−0.29228	38.504	11.41
4	−0.52303	30.048	19.85
3	−0.75378	22.549	27.34
5	−0.29228	38.504	11.41
4	−0.52303	30.048	19.85
6	−0.06153	47.547	2.4
6	−0.06153	47.547	2.4
4	−0.52303	30.048	19.85
7	0.169216	56.719	6.36
3	−0.75378	22.549	27.34
21	3.399713	99.966	49.97
7	0.169216	56.719	6.36
6	−0.06153	47.547	2.4
5	−0.29228	38.504	11.41
8	0.399966	65.541	15.54

The way this works with the normal curve is in terms of probability density. Those *z* scores around the mean, in the middle of the bell, are more likely and so have a higher probability of occurring under normal circumstances. Those scores further out in either tail have a much lower probability of occurring. Because of the nature of the normal curve, we will look at the percentage of area under the curve that sits below or above the point of interest, and *that percentage equates to the probability of that outcome occurring.*

Let's add a column or two to Table 5.4: To get the percentiles in the third column, I used Microsoft Excel's "NORM.S.DIST" function.

Looking at the second and third columns in our table, we can see that for our distribution, a raw score of 3 falls a little over three quarters of a standard deviation below the mean (−0.75378). That lands it just above the 22nd percentile. Our score of 8 is almost 0.4 of a standard deviation above the mean, which places it just above the 65th percentile. The raw score of 21, the one that interests us, that score is nearly 3.4 standard deviations above the mean. How likely is it to happen normally? As it is in the 99.966th percentile, it is *exceedingly* unlikely, happening less than 1 time in 1,000 under normal circumstances.

For the last column, I used the normal distribution table in the back of this book found in Appendix A. Values for areas under the curve (or AUC) typically are presented in one of two ways. They can denote either the AUC from the mean to the *z* score, or they can present the AUC from the *z* score to the far end of the distribution (because *z* is symmetrical, the denoting of AUC via either method works the same regardless of the sign of the *z* score). In our case, for Appendix A, the table denotes the distance between the mean and the *z* score. If you look at the raw scores = 6, the *z* score is small, only around −0.06153*z*, which is very near 0 (or the mean). It stands to reason, then, that the space between the mean and that score is only a tiny portion of the area under the curve. It is tiny indeed; that wedge of probability territory is only about 2.4% of the total space under the curve. For our 21, however, it has a *whopping z* score, nearly 3.4*z*. That's huge, so it stands to reason that the space under the curve from the mean to that 21 is nearly the entire half – 49.97% to be exact. AND, if we think of that score in terms of the rest of the scores, we should remember that half the distribution lies below the mean, and 21 is well above them, too. That means that our 21 score falls at or near the 99.97th percentile (we add the 50% below the mean to the number from Appendix A to get the total distance from the bottom of the distribution to that point).

Now that we can approximate the relative likelihood of a given outcome using the context provided by the normal curve (as an approximation of the binomial distribution), we are one big step closer to being able to make some decisions using data.

Summary

In this chapter, we took some rather huge steps toward a far grander statistical capacity: we began by learning about basic probability principles and then ramped them up into the binomial theorem, but we weren't done yet. That bell-shaped arc, the one that emerged from probability and multiple coin flips, is approximated by the Gaussian, or bell curve, and that probability density function allows us the capacity to take z scores and turn them into relative probabilities. So equipped, we can contemplate outcomes of myriad sorts and determine their relative likelihood under "normal" conditions. Chapter 6 will equip us with the ability to put what we've come to understand here to work in the human sciences.

Main Take-Aways

- The scientific interpretation of quantitative data requires of us that we derive some form of context for comparisons. As substantive meaningfulness is beyond the highest capacity of statistics, we can derive context in other ways. Often (but not always), statistical analysis utilizes probability theory to provide the context.
- Probability theory is a way to speak to the likelihood of an event occurring under conditions of randomness. Randomness isn't about reckless abandon or disconnection from conditions. Indeed, randomness implies connection and interdependence.
- The Bernoulli experiment and the binomial distribution that results from it serves as a foundation for the creation of a context that can serve us ably.
- The normal curve – bell-shaped, symmetrical and derived through an equation – allows for the capacity to talk about the relative likelihood of any given outcome, and to speak to that likelihood in precise, probabilistic terms.

Key Terms (in order of appearance)

population
sample
experiment
event
degrees
sample space
probability
relative frequency
complement
replacement/non-replacement
addition rule
multiplication rule
mutually exclusive
binomial distribution
Bernoulli experiment
observed outcomes
expected outcomes
probability density function

Chapter 6

The Central Limit Theorem and Sampling

Here's the punch line: the set of scores before us in our spreadsheet or database (whether they be from a sample or a population) constitute our empirical data from which we can make an empirical distribution of outcomes (like our *actual* coin flips from Chapter 5). Then we can select a probability density function, like the Bernoulli or z probability density functions, to provide a measure of context for the empirical data. The probability density function constitutes the theoretical distribution that provides the *context* for understanding the meaningfulness of the scores (and also quite a bit more).

Being able to approximate the relative likelihood of a given outcome is a useful strategy for creating context. It certainly isn't the only strategy to provide context, and it probably isn't the best strategy (except in some instances), but as a strategy, it has some marked advantages. First, the context it provides is particular to the setting of interest – we can talk about the relative likelihood of getting a particular score given a sample where the mean and standard deviation are known (that's exactly what z scores are). And because there are actually shedloads of probability density functions, we can fine-tune our probabilistic framework and assumptions to fit our needs for the moment. Second, context provided by probability density can, under certain circumstances, be independent of other kinds of contexts – we don't have to know any more about what the scores actually measure. Third, the context provided by probability density functions is comparable across a wide variety of events in the human world. Probabilistic logic, though flawed, is both satisfactorily meaningful and comparable across settings.

Of course, using probability density functions for context also has some drawbacks. First, as above, the substantive meaning of a set of scores can be taken to be incidental or less important in ways. We will talk later about the meanings of the word, "significance." For current purposes, it is sufficient to know that significance (for example, the meaningfulness of a score or comparison) can be lost (or at least become blurry) when we use **statistical significance** as an understudy for operational significance (e.g., Ziliak & McCloskey, 2011).

Descriptive and Inferential Statistics

Now, there is a key distinction we need to make at this point. Statistics, like the measures of centrality and dispersion we have discussed so far, are called "descriptive." **Descriptive statistics**, well ..., *describe*, or depict important aspects of a variable to impart a larger sense of the variable. Like describing a car in terms of its size, color and trim, centrality, dispersion as well as skew and kurtosis are ways of characterizing a variable. The particulars of a variable, say, the size of the standard deviation or the difference between its mean and median,

DOI: 10.4324/9781032638591-6

definitely interest us when it comes to using that variable in analyses with other variables (bivariate or multivariate). Precisely why that is the case is a thing that must wait for later.

In contrast to descriptive statistics, and importantly one reason statistics can be so helpful, is another kind of statistics. **Inferential statistics** are so called, really for two reasons. The main meaning of the term "inferential" statistics is in relation to their use: inferential statistics allow us to make inferences about the world (about data we don't have) based on patterns in the data we *do* have. It is also the case, though, that to allow for inference-making, certain assumptions (or inferences) must be made about the data used in inferential statistical techniques. In later chapters, we will talk about *t* tests, ANOVA, correlation and linear regression. These analytical techniques are examples of inferential statistical techniques.

Reasoning toward Inference

There is a bit of reasoning that holds inferential statistics together, reasoning that allows us to infer things about the wider world from data in our possession that do not directly measure the whole of the world. This bit of reasoning is related to the Bernoulli experiment and the probability density functions discussed in the previous chapter. Recall that the probability of any given single outcome, say, on a fair six-sided die, was the same as any other outcome. Table 5.1 presents probabilities for five flips of a fair coin, and the probability of getting any given number of heads is a function of the probability of any one five-flip outcome (which does not change) and the number of ways an outcome (say, three heads) could occur. The probability of success didn't change (because each flip is an independent event), but the number of ways success can happen does change. Zero heads on five flips requires a rather particular set of circumstances; in fact, there is only one way that such an event can occur. The same is true with getting five heads. There are, however, rather more ways (i.e., combinations of heads and tails) for three heads to result.

In essence, what we have done with repeated flips is to repeatedly "sample" from a "population." Recall that the probability of any given outcome is shown in the back half of the equation (in bold below):

$$P(x) = \frac{n!}{r!(n-r)!}\boldsymbol{p^r q^{n-r}}$$

As the probabilities of success and failure change (again, say, the probability of three heads on five flips versus the probability of getting two heads on five flips), the actual probability term in the equation itself does not. Again, it can be understood as the independent probability of deriving one success given the number of trials, and again, we can think of it operating like a weight. Again, the binomial probability formula above is a more sophisticated special application of the addition rule of probability.

This point about equal probability is *absolutely vital*: in most of what's ahead in this book, we will operate with the assumption that *the probability of selection of any particular member of a population is the same* for every member of the population. Oftentimes when we teach human sciences research methods we talk about the classical experimental design (CED), and we treat it as something of a gold standard (which I won't do here). One of the core assumptions of the CED is the idea of equal probability of selection (again, sometimes called EPSEM). The reasoning that allows us to infer things about a population on the basis of a sample drawn from that population is founded on the equal probability of selection

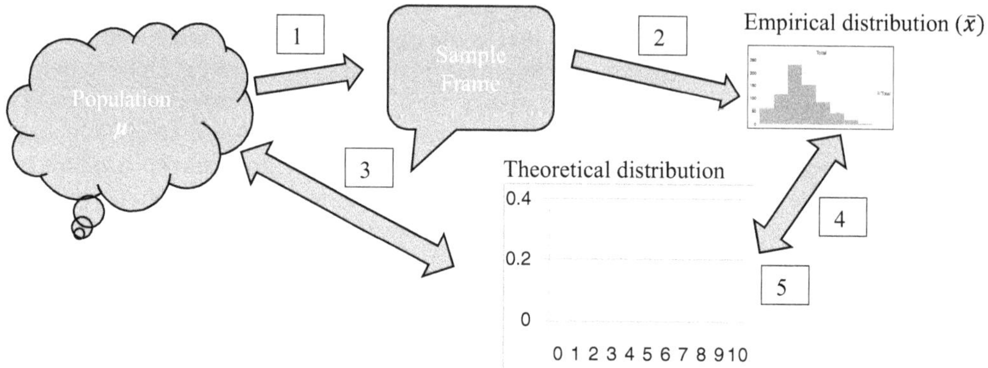

Figure 6.1 The Cast of Characters and Their Relationships.

assumption (that each individual outcome – like a coin flip – has the same probability as any other, just as we see in the Bernoulli experiments in Chapter 5). The equal probability assumption maximizes the representativeness of population identity for any given member of that population.

Let's consider the setting so far (see Figure 6.1). We have a population (or a sample frame standing in for a population) about which we want to know things. Specifically, we want to know a value of a variable in the population – we call this value μ – *and it is a thing we will never know directly*. We also have an empirical distribution of some kind, some real-world data drawn from a sample from that sample frame. We want to be able to infer things about the population (μ) on the basis of what we know about that sample, and so we need some kind of context in order to do so. Comparisons work for context, but if we could make a comparison with the population, we wouldn't need to draw a sample in the first place. To stand in for the missing details from the population, we have a theoretical distribution, one of many, in fact, from which we could pick, that gives us a reasonable probabilistic proxy for things happening in the population (or in the sample frame).

1 The population is the idea, the thing about which we are curious. We will not ever know details about the population (μ) sufficient to create a distribution of its scores. A population is a changeable, amorphous thing with permeable boundaries, its volatility rendering precision nearly impossible. Instead, we achieve precision by establishing a *sample frame* that defines in discrete terms a set of measurable constructs (like people or events) that we can state in a definite way.
2 The sample frame, standing in for the population, often is still too large to measure in totality, so we sample from it. The sample we draw will give us our empirical distribution.
3 But having the sample and its distributions does not tell us much about the population. We need some kind of theoretical distribution against which to compare our sample terms. We don't know parameters of the population (like the population mean, μ); if we did, we wouldn't have to sample in the first place. So, to stand in for what we don't have in either the population or the sample frame, we use a theoretical distribution (e.g., z, t, F...).
4 For that sample to be of any value in analyses, we need to know something about how it was drawn. For our purposes here, a random sample, one where every member of the

sample frame has an equal probability of selection, is optimal (interestingly, it isn't necessary if we judge by common practice, and also by some rather convincing evidence, even though it is preferred). If that sample is a sample of my favorites, or of all sample frame members that begin with the letter H, though, it ceases to be random, and its ability to support generalization is attenuated.

5 The theoretical distribution, acting as stand-in for the population (much like the sample frame is doing in its own way), provides context for comparisons with which to consider the sample for the purpose of making inferences about the population.

Because we can't compare the sample to the population directly (because what we want to learn about is the population itself), we need to be able to compare the empirical distribution with the theoretical one(s) – we need to have a bit of reasoning that allows us to walk back and forth between the empirical and the theoretical distribution(s) without stumbling. It turns out the reasoning is pretty straightforward.

Sampling Distribution of the Mean

Here's that bit of reasoning I have been promising. We can take multiple samples from a population (*with replacement*, most likely – remember replacement from earlier?), just as you reasoned we might do just a moment ago. Then we can take the mean of each of those samples and with *them* make a whole new distribution of scores. Our new distribution composed of means, one mean from each of the samples from the population (or the sample frame) is a **sampling distribution**. And here's a bit of magic: the shape of the sampling distribution will be normal (or bell-shaped) *regardless of the shape of the empirical distribution.* That sounds nuts, right? Kachigan (1986: 106–107) provided what might be the clearest demonstration of why this is, so we will employ it here. Check this out.

We have a population of four scores (1, 2, 3 and 4), each with an equal probability of arising (0.25). Figure 6.2 shows the probabilities for each of the possible outcomes (sampling is done with replacement).

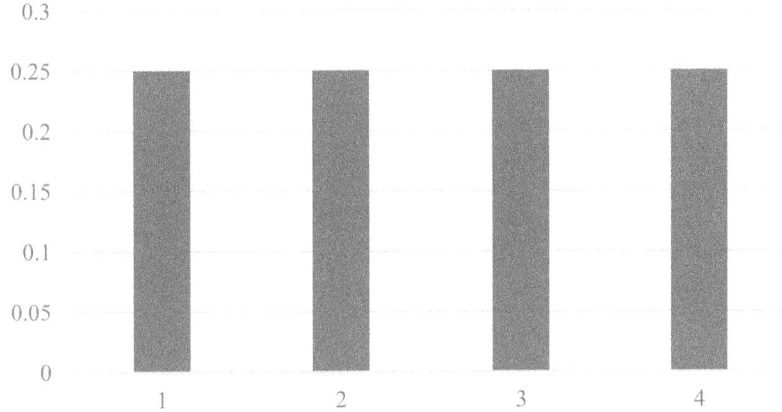

Figure 6.2 Probability of Outcomes, Parent Population.

So, in order to discover the mean of the population by using the mean of the sample as a best guess, let's say that we sample one case from the population, replace it, and then we sample another case, and then take the mean of the two values we draw. The outcomes would look like this:

Notice in Table 6.1 that of the sixteen possible outcomes for $n=2$, the "correct" mean ($10/4=2.5$) is possible four times (drawing 1 and 4, 4 and 1, 2 and 3, or 3 and 2). The probability of a sample of two cases from a population of four cases deriving the "correct" mean is 4/16, or 0.25. There are three ways in which a mean of 2.0 can occur, just as there are for a mean of 3.0 to occur; the probability of getting a mean of either 2 or a mean of 3 (not in a conditional probability sense) is 3/16 or 0.1875. If sampling is random (that is, systematically non-systematic), imprecise estimates become increasingly improbable. Precision increases, of course, in direct relation to the size of samples drawn randomly.

The probability density histogram for this table looks like this (Figure 6.3):

Again, the probability of sampling into the "correct" mean is 4/16, or 0.25. That means that any given pair of two samples drawn from the population of four scores (1, 2, 3, or 4)

Table 6.1 Outcomes for sampling $n = 2$ and corresponding means

	Second Score			
First Score	*1*	*2*	*3*	*4*
1	(1 + 1)/2 = 1.0	(1 + 2)/2 = 1.5	(1 + 3)/2 = 2.0	(1 + 4)/2 = 2.5
2	(2 + 1)/2 = 1.5	(2 + 2)/2 = 2.0	(2 + 3)/2 = 2.5	(2 + 4)/2 = 3
3	(3 + 1)/2 = 2.0	(3 + 2)/2 = 2.5	(3 + 3)/2 = 3.0	(3 + 4)/2 = 3.5
4	(4 + 1)/2 = 2.5	(4 + 2)/2 = 3.0	(4 + 3)/2 = 3.5	(4 + 4)/2 = 4

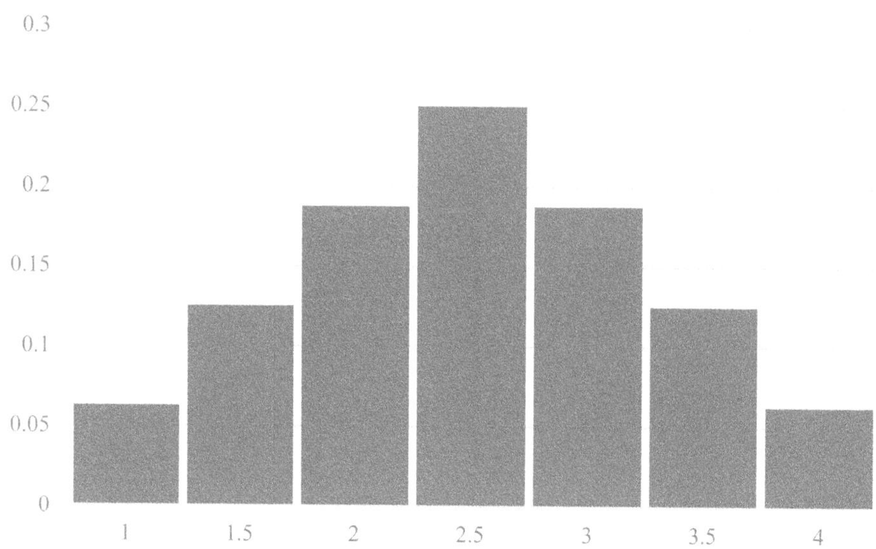

Figure 6.3 Sampling Distribution of the Mean, (Four Equally Likely Scores, $n = 2$).

has a 0.25 chance of being "correct." Notice also that the shape of the sampling distribution is solidly normal (i.e., Gaussian or bell-shaped, just like the binomial distribution for a fair coin) whereas the actual parent distribution shown in Figure 6.2 is *decidedly* not (see Figure 6.2).

You're probably a practical sort, right? It serves you well. Well, let's put that practicality to work here. Let's say you want to discover the mean value (μ) of a variable within a population. Let's also say that each possible value in that population has an equal likelihood of occurring. Further, let's say that you had to draw a tiny sample (say, $n = 30$ from a population of 100,000). How comfortable would you be that the mean of the sample ($n = 30$) is *exactly* the same as the population mean? Not too comfortable, right? Good to hear. You might be sort of comfortable in saying that it is *somewhere near* the actual population parameter, but that's not the same thing, now, is it? We will talk about confidence intervals soon. No, if you want to be more certain about the quality of your estimate of the population mean, you will want more samples, right? You'd go back for more information, right? Well, that's precisely what we would do, too.

If we take repeated samples (again, with replacement, at least theoretically) from a population, calculate the mean of a variable that interests us for each sample, and then treat each mean as a new value in a new distribution, we have cooked a sampling distribution. As we have seen already, regardless of the shape of the population, if we sample enough, the sampling distribution of *r* samples from that population will be normal. *What we haven't seen yet, though, is that as r (the number of samples) increases, the sampling distribution will become increasingly peaked, or leptokurtic.* That seems nuts until one recognizes that the sampling distribution is seeking to portray the *mean* of the variable in population, not the *shape* of the population itself. *What is more, though, is that the mean of the sampling distribution, that mean of the means, is our best guess at μ – that thing we crave knowing.*

Here's the thing, though: we can't always go back in for more samples or go back and draw a bigger single sample (which amounts to the same thing, sort of). To make things worse, we might not know the actual size of the population (or even the size of the sample frame), and so any measure of the mean of that population is going to be an estimate. And it gets worse from there; human sciences data are notoriously buggy, lumpy and imperfect, and one of the continual challenges in human sciences research is navigating between the ideal conditions for which statistical analyses were designed and the weird and wacky world of real data.

If repeated samples are the best bet statistically, but such a thing is rarely in the cards logistically, what do we do? What we need to be able to do, then, is to assess the quality of the one guess we have in the body of the sample mean before us relative to the population value μ.

Central Limit Theorem

It turns out we aren't sunk. Not at all. Remember that the sampling distribution is a distribution like any other distribution; it has a mean, of course (and again, the mean of the means – the mean of the sampling distribution – is our best guess at the population parameter, μ). That sampling distribution also has a variance, too. Because we can calculate a mean, we can compare that mean to each score (recall that in a sampling distribution, each score is actually a mean) in order to cook a standard deviation. Well, that standard deviation is magic for us; if we know the standard deviation of the sampling distribution, we *don't have* to sample repeatedly, because we will always have a sound sense of the quality of any given sample we draw. In fact, the standard deviation of the sampling distribution (which varies inversely with n – it gets smaller as sample size gets bigger) has a special name: **the standard error.**

For some of the same reasons that we can't simply measure a population's variable values, neither can we actually *know* the standard error for any given sample (usually). The good news, though, is that we know how to estimate it. We call this the **standard error of the estimate**, and it is derived using the formula:

$$\sigma_{\bar{x}} = \frac{\sigma}{\sqrt{n}}$$

Or said in English, the standard error of the estimate (our approximation of the standard deviation of the sampling distribution) is equal to the sample standard deviation divided by the square root of the sample size.

One can now see why standard error varies as a function of sample size: look back at the example shown in Table 6.1. As sample size increases, it becomes increasingly unlikely to draw a sample from that population that does not approximate, or resemble, the mean of that population. The sampling distribution, then, becomes increasingly normal and leptokurtic as sample size increases. Thus, we have reasoned our way to the Central Limit Theorem.

The Central Limit Theorem states:

> Given a randomly drawn variable from a population, the sampling distribution of a population will become increasingly normal (that is to say, bell-shaped) as sample size increases, regardless of the distributional shape of the parent variable in the population.

Again, we have reasoned our way to the core logic that holds together a large portion of inferential statistics. Because of the Central Limit Theorem, we are on stable ground when we infer things about a population based on measurements taken from a sample, provided that sample meets some key assumptions:

- Sampling from the population is random. Another way to say the same thing is that each member of the population must have an equal probability of selection;
- Each sample should be independent (e.g., sampling with replacement);
- The quality of the estimate of μ varies with the sample size (we will talk more about the qualities of our estimates, sample size, statistical power and such in Chapter 7), so sample size must be sufficient (an $n = 30$ is thought to be "large" enough);
- The variable of interest for μ should be continuous or metric (because it needs to have a meaningful mean and standard deviation).

Let's consider a few examples to enlarge on what we've discovered.

Recall that the Police Data Initiative (www.policedatainitiative.org) provides unfettered access to police agency data. We will begin with a look at the Rochester, NY Police

Department's employee spreadsheet as of 6/23/2020 (the spreadsheet is included with this package – it is called "RPD_Police_Personnel"). The spreadsheet presents information for 896 employees, both sworn and nonsworn. These data will stand in for our population. In this instance, because we can measure μ directly, we should be able to witness some of the things we have been discussing with real data. Specifically, we will be able to see what a sampling distribution from a parent population looks like in actuality. We will also be able to see how sample size and random selection impact the quality of our estimate of μ.

For current purposes, let's take a look at employee age at hire. The youngest person at point-of-hire currently in the employee data is 22 years old; the oldest is 69. The mean age at hire is around 27.93, with a standard deviation of about 6.67 years. The mean, 27.93 stands in as our population parameter, our μ. As can be seen in Figure 6.4, it turns out that the empirical frequency distribution of age at hire is not exactly normal, just as one might expect. It has a notable positive skew with a rather long tail rambling off to the right. Sworn staff tend to be relatively young at hire.

So, let's say we don't have this information, but that we need to approximate the population parameter of mean age at hire. Using Microsoft Excel's random number generator, I created a sample, $n=50$, which amounts to about 5.57% of the population. That first sample looks like this (Figure 6.5):

The mean for this sample was 28.755 and the standard deviation was 7.34. The estimate of μ from the sample, that is to say the sample mean, was only about 3% away from the population parameter, which isn't too bad. The sample histogram resembles the population in some very general ways (for example, the positive skew evident in the population is

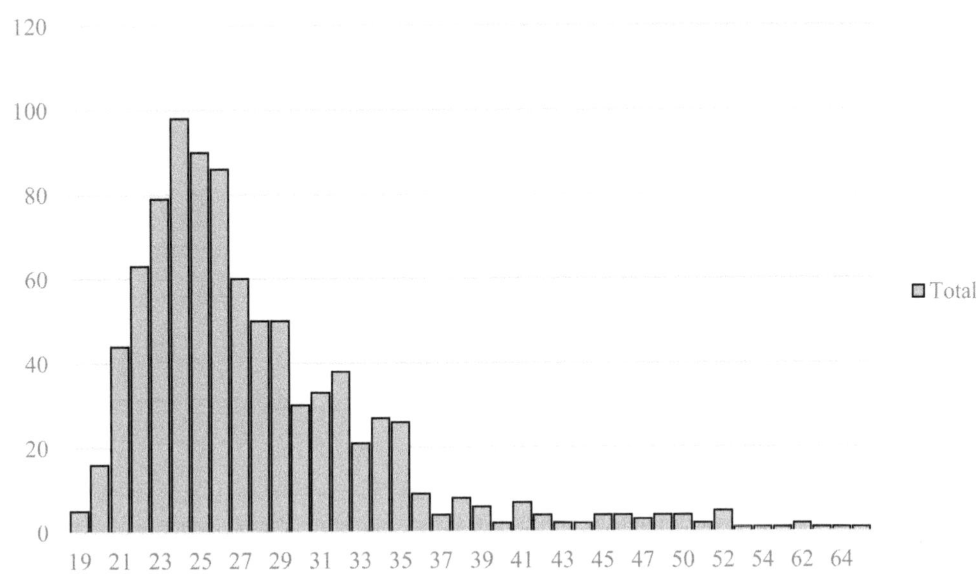

Figure 6.4 Frequency of Age at Hire, Rochester Police Department, 6/23/2020.

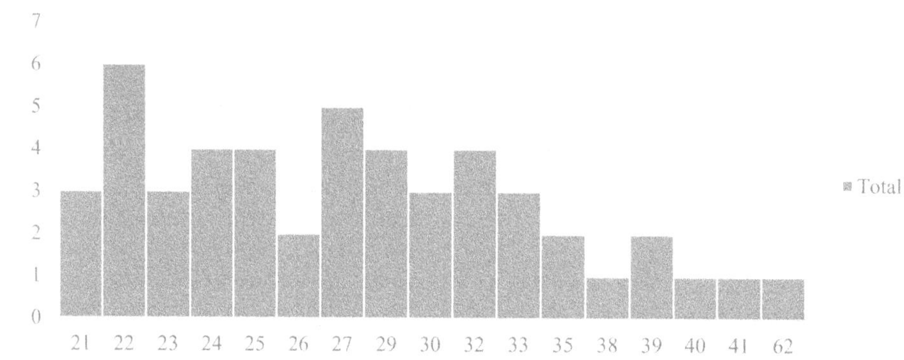

Figure 6.5 Our First Sample from the Population, Age at Hire ($n = 50$).

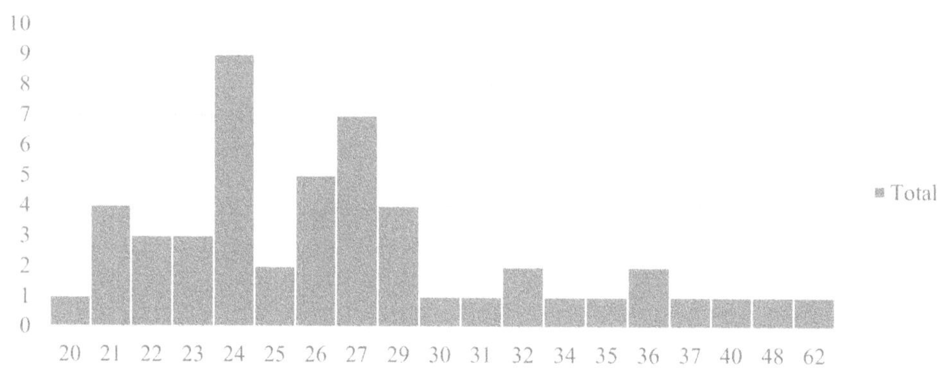

Figure 6.6 Our Second Sample from the Population, Age at Hire ($n = 50$).

Table 6.2 Standard Error Estimation

Sample	*N*	*Mean*	*Standard Deviation*	*Standard Error*
First	50	28.755	7.341	1.038
Second	50	27.86	7.387	1.045
Actual Standard Deviation of the Sampling Distribution (n = 2)				0.633

somewhat present in the sample, and the overall shape is comparable). Still, we should not rest on laurels. The second independent sample looks like this (Figure 6.6):

For our second sample, also drawn via random number generation ($n = 50$) after replacement, the mean was 27.86 and the standard deviation was 7.387. In all, the second sample constitutes a more precise guess at μ, but of course we would have no way of knowing that were we to be without a population to which to compare.

Now if we sought to have the best possible guess at μ (in this case the mean age at hire for the population), we would create a sampling distribution ($n=2$) and find its mean. Again, the formula for a mean is:

$$\bar{x} = \frac{\sum x}{n}$$

The first sample mean (28.755) plus the second sample mean (27.86) is 56.615 (recall that there are just two scores to add together). 56.615/2 = 28.31. That is the closest estimate yet to the actual μ of 27.93.

Now check this out: because we actually made a small sampling distribution, we can actually find its standard deviation. The standard deviation for the sampling distribution ($n=2$), or the standard error, is about 0.633.

Recall that we can estimate the standard error (the standard deviation of the sampling distribution) using:

$$\sigma_{\bar{x}} = \frac{\sigma}{\sqrt{n}}$$

If we were to estimate the standard error based on either of the samples, we would get:

The estimates of the standard deviation of the sampling distribution (the 1.038 and the 1.045) were a little conservative and overestimated the amount of variance, but clearly for the calculation, the larger the n, the better an estimate one derives. If, say, the standard deviation for the first sample was the same, but it had been based on an n of 70 cases, the square root of n would have been 8.37 or thereabouts, increasing the size of the denominator relative to the numerator, and thus making the result a decimal rather than a whole number (specifically, the estimate of standard error would have been around 0.877), considerably closer to the real value.

In Chapter 7, we will focus the balance of time talking about the quality of estimates and about statistical power. We will return to this reasoning in Chapter 10 when we talk about hypothesis testing for independence.

Summary

We have crossed the Rubicon into inference: the ability to infer things not in direct evidence about the human world. We discussed sampling briefly again, a conversation that is likely to be familiar from your methods class. But the why behind sampling proclivities might well be clearer now, given our discussion about sampling distributions in support of inference. Then there's the Central Limit Theorem, the keystone upon which Ordinary Least Squares inference depends.

Main Take-Aways

- Some statistical techniques, like the mean and the standard deviation, operate in a way that merely describes a set of data. Other techniques, like those we will discuss in future chapters, allow us to infer things about a population on the basis of the sample we have before us.

- We can reason our way to being able to deduce things about a population on the basis of information contained in a sample when certain things about that sample are true.
- In order to determine the parameter of a population, we can take repeated samples, merge those samples into another set of scores called a sampling distribution; the sampling distribution enables reasoning about the parent population.
- The Central Limit Theorem, bringing the above together, allows samples to reliably give us useful information about a population: when samples are random and independent, and of adequate size, the sampling distribution will be normal regardless of the shape of the empirical distribution. Further, the mean of the sampling distribution is our best approximation of the population parameter μ.

Key Terms (in order of appearance)

statistical significance	μ (Mu)	standard error
descriptive statistics	sampling distribution	Standard error of the estimate
inferential statistics	Central Limit Theorem	

Chapter 7

Point Estimates and Confidence Intervals

We might wonder whether there is some measure of melancholy for astronomers and astrophysicists. After all, at least for those who study distant objects like other stars, there is absolutely no hope that they will ever actually visit the objects they study. They must find contentment with things as we can learn about them standing stuck on our terrestrial home.

The analogy of astrophysics to point estimates is illustrative of a crucial point: we want to *know or experience μ* but *won't ever actually see it*. Instead, we must be content with reasoned approximations. Sometimes those approximations can serve most admirably; after all, sampling a distant star's light gives a measurable sense of the star's chemical contents and velocity (i.e., redshift) in some detail. It is, in a rather precise sense, a systematically nonsystematic sample (oddly, we might say that it is nonsystematic from the star's perspective and at the same time a convenience sample from our own).

We can think of inferential statistics in terms of three distinct but related kinds. The first kind of inference, point estimation, relates to the contemplation of a sample estimate $\bar{x}$ of a population parameter μ. The second kind of inference, quite deeply related to the first kind, is about the estimation of intervals of confidence around that point. The third kind, hypothesis testing, will be the focus of a considerable portion of the remainder of this book.

Point Estimates

In truth, we have already begun thinking in terms of the assessment of point estimates. We just made and assessed point estimates at the end of Chapter 6. We attempted to estimate a value in a population, μ, on the basis of some sample statistic $\bar{x}$. Judging from our results, we did rather well.

Again, though, we should not rest on laurels and assume a sample of size n will always score us a scientifically responsible estimate of μ. Pretty clearly, not all samples are created equal. We might think about point estimates, our estimations of μ based on $\bar{x}$ in terms of (a) bias and (b) parsimony (e.g., Kachigan, 1986). If we seek to generalize results from a sample to a population, we want to know that the sample is (a) unbiased – that the mean of the sampling distribution is a close estimate of μ and (b) efficient – that it brings us to a precise estimate of μ in an efficacious manner. The mean and the median are both unbiased estimators of μ, but a sampling distribution of means likely will have a smaller standard error than a sampling distribution of medians of the same sample size. Thus, the mean is a more efficient estimator of μ. In essence, then, *we want a sample to be as representative of the breadth and depth of the innate variance present in the population as possible, and we*

DOI: 10.4324/9781032638591-7

want our measures of that sample to be able to use as much information in that sample as can be managed.

Probability Distributions and Inference

Recall in Chapter 5 how we talked about the binomial distribution? Well, that bell curve derived from probabilistic reasoning is one of several (actually, many) theoretical distributions that we can use for comparison purposes in order to support our three kinds of inference. How, precisely?

It turns out that these theoretically or mathematically derived probability density functions, like the binomial distribution we created earlier with the coin flips, what we call the normal (Gaussian or *z*) distribution, any of the *t* or *F*, distributions, or the chi-square (X^2) distribution, just to begin, have known and precise shapes. These shapes are so precise, in fact, that we can speak in very detailed terms about the area under a probability density curve, which is the same thing as speaking about how relatively likely a given outcome is.

Figure 7.1 (which we saw earlier in Chapter 5, Figure 5.3) shows that there are hunks of the normal curve that represent knowable and precisely expressible portions of probabilistic real estate, the property lines being chunked out by the natural longitude lines created by sample standard deviations. More particularly, what we see in Figure 7.1 is what we call a standardized or *z* distribution. *Z* distributions like this one have a mean of 0 and a standard deviation of 1.0, as we have established already. Now what that allows us to be able to do is to talk about those hunks of real estate in very precise terms. The area under the curve between the mean (0) and one standard deviation above the mean is 34.13%. That means that of all the area under the curve here, 34.13% of the total area (and presumably, 34.13% of all variable values) falls between 0 (the mean) and one standard deviation above the mean.

The same is true of the space between the mean and one standard deviation *below* the mean. Pretty cool, yeah? It gets better. If we wanted to talk about the space under the curve between one standard deviation below the mean and one standard deviation above the mean, we can add those two numbers together: 68.26% of the area under the curve falls

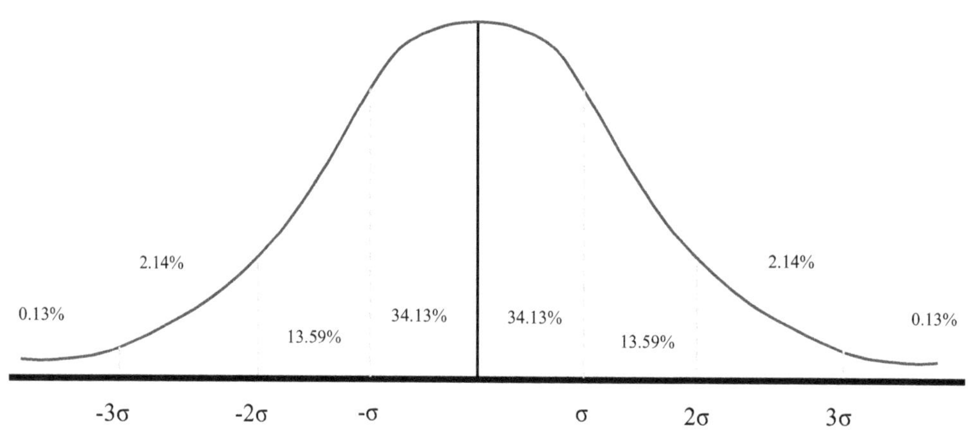

Figure 7.1 Area under the Normal Curve.

between -1σ and 1σ. That means that we can expect that nearly 70% of scores of, say, shoe sizes, IQ scores or police employees' age at hire, will fall between -1σ and 1σ. Notice also that half of the distribution (or very nearly half) lies to either side of the mean; the distribution is symmetrical so both halves are even, and if you look at both Figures 7.1 and 7.2, you'll notice that the tails never quite touch the bottom line, so the probabilities never actually reach 0.0. We call this property of never having a 0.0% likelihood **asymptotic.** An asymptotic probability curve never expresses a point of absolute impossibility.

Looking further away from the mean, we can see that the space between 1σ and 2σ above (or below) the mean corresponds to 13.59% of the area under the curve; 13.59% of scores, then, *if no other forces are working on the data (and this is key and becomes more crucial as we go on)*, can be expected to fall in the space between 1σ and 2σ. We could also say that nearly but not exactly half of the distribution (34.13 + 13.59 = 47.72%) falls between the mean and two standard deviations (2σ) above (or below) the mean (see Figure 7.2). That also means that 95.44% of the area under the curve falls between -2σ and 2σ. If we go further and include three standard deviations, plus and minus, we end up with more than 99% of the area under the curve falling there (meaning that more than 99% of a distribution's scores *should reasonably be expected* to fall between three standard deviations above and below the mean.

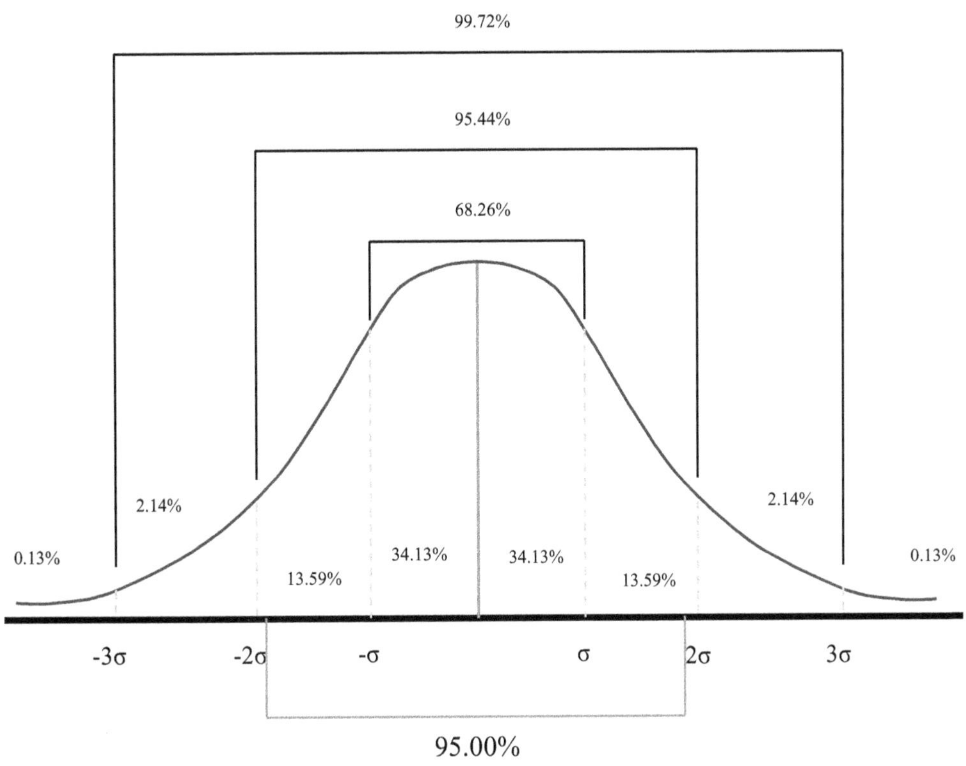

Figure 7.2 Areas under the Curve between $-K\sigma$ and $K\sigma$.

Of course, we also work backward, too, here: if we wanted to hunk out a portion of the curve, say, 95% of it between the lowest score and the highest, we could add the chunks together to get a number of standard deviations. Figure 7.2 demonstrates how we might apply these insights regarding area under probability curves to expectations. If nothing else is influencing our sample (meaning that it is an unbiased member of the unbiased sampling distribution that ably approximates μ), then we can say with some confidence that any given sample has a knowable likelihood of falling in a given place on that curve. It turns out that if we wanted to deal with exactly 95.0% of the distribution arranged symmetrically around the mean, we could use 1.96σ as our boundaries: the space between -1.96σ and 1.96σ is exactly 95% of the area under the curve.

We have a sample of size n, and that sample has a mean, $\bar{x}$. We can calculate a standard deviation for that sample and thus we can approximate the standard error of the (hypothetical) sampling distribution from which it has come (again, called the *standard error of the estimate*). We have reasoned that the sampling distribution is normal and symmetric in shape, with knowable probability densities (the area under the curve). That all means, then, that we can assess $\bar{x}$ relative to the sampling distribution using the normal distribution probabilities.

If we assume that a sample is an unbiased and efficient estimate of a population parameter, we are also assuming that the sampling distribution to which that sample belongs has a particular shape (often bell-shaped, meaning either z or a t distribution, although there are others, like X^2 and Poisson that are not bell-shaped). This bit is key: we want our *sample* to resemble the population, but we don't assume that our *sampling distribution* looks like the population distribution. Because of the Central Limit Theorem, we assume, in fact, that the sampling distribution resembles a normal curve. And that normal curve, because it is so closely mapped, gives us a sense of the relative likelihood of a given outcome (i.e., the mean of a sample, $\bar{x}$, being representative of μ). Don't forget: the sampling distribution operates to tune in on the population *parameter*, not the shape of the population *phenomenon*.

Assuming the sample we have is one of many possible unbiased sample representations of the population of interest and is of sufficient size, a hypothetical sampling distribution from those many possible samples would be normal in shape, with a mean of the sampling distribution being our best guess at the population parameter (μ).

OK, so we can infer the relative likelihood of an unbiased sample score x occurring within a sample. Recall the business about z scores: they give us a standardized expression of a given outcome for a case relative to the sample mean and standard deviation. We can also talk about a sample statistic $\bar{x}$ in terms of the approximate standard error of the sampling distribution from which it could have come. The z score provides us a way of speaking to relative likelihood that we will be using, but the point estimation refers to the reasoning that allows us to use $\bar{x}$ as an approximation of μ.

Confidence Intervals

So now we have a *point estimate* for μ in the body of our sample mean, $\bar{x}$, but again much is not known relative to the quality of that estimate. Don't forget that we are making assumptions about our sample relative to the sampling distribution's shape – a shape we can't confirm – all in order to estimate a population parameter that we will not ever actually know. Perhaps it is more precise to say that given what we know about $\bar{x}$, we might be on firmer footing if, in addition to providing our best guess at μ, we also provided something of a window of error. We might call it a **margin of error.** I dig that. Actually, no doubt you've heard that phrase, "margin of error" used before, right? Polling firms present poll results, say, regarding political candidates or policy ideas, with plus or minus three-point margin of error and, say, 95% confidence. Well, that means that their best guess at the real value μ (the thing we can't actually know but hope to know something about) is their sample mean $\bar{x}$, and that they are 95% certain that the actual value of μ exists within a window of 3% points on either side of $\bar{x}$. This margin of error gives a window within which one might be reasonably certain, say 95% certain, that the actual population parameter value is the mean.

The reasoning looks like this:

$$\mu \approx \bar{x} \pm z(\sigma_{\bar{x}})$$

This says that our best guess at the population parameter μ is the sample mean ($\bar{x}$) plus or minus the confidence level (z) times the standard error of the estimate. Now, we can think of the $z(\sigma_{\bar{x}})$ expression as being our **margin of error,** so the whole thing simplifies to

$$\mu \approx \bar{x} \pm \text{margin of error}$$

In order to create our confidence interval, we need a few things. First, of course, we need the sample mean. Then we need to decide about our level of confidence. Typically, we tend to choose 95% confidence as our acceptable standard. That means that being 95% confident about the sample mean relative to the population is sufficient. **That means also that one sample in 20 from a population – one sample in 20 from our sampling distribution – will have a mean that is naturally nontrivially different.** If we are going to follow the 95% confidence convention, we would use 1.96 as our value for z. Finally, we need our standard error. Recall that we can calculate the standard error of the estimate:

$$\sigma_{\bar{x}} = \frac{\sigma}{\sqrt{n}}$$

Or our estimate of standard error – that is, the standard deviation of the sampling distribution, is the sample standard deviation divided by the square root of the sample size.

If we wanted to calculate a 95% confidence interval for our sample mean, we would use:

$$\text{CI} \approx \bar{x} \pm 1.96(\sigma_{\bar{x}})$$

Let's take an example. Say we have a mean of a sample ($n=40$); let's make the mean 13.2. That same sample has a standard deviation of 3.61. We have all we need to make a 95% confidence interval – we can say through that confidence interval that, based on the sample, the population parameter has a 95% likelihood of being either the mean on the button or within the confidence interval.

OK, so first, we need to calculate the standard error of the estimate. The sample size is 40; the square root of 40 is 6.325 (rounded up). Our standard error of the estimate, then, is $3.61/6.325 = 0.571$ (rounded up). The heavy lifting is already done.

Next, we can multiply 1.96 times our standard error of the estimate to get our margin of error. $1.96(0.571) = 1.119$, and that's our margin of error. Our mean was 13.2. For our confidence interval, then, we first add the margin of error to the mean and get 14.319. Then we subtract the margin of error from the mean and get 12.081.

$$\text{CI}_{95} \approx 13.2 \pm 1.119$$

Or another way to say this is that we think that the population parameter is the sample mean (13.2) but we are 95% confident that it falls between 12.081 and 14.319.

Let's work another example. As we mentioned earlier, your police department performed a community satisfaction survey recently. A sample of individuals in the community was selected using as nonsystematic a strategy as possible. Four questions measured respondents' feelings about police officers, one question each invited respondents to grade officers' professionalism, responsiveness, helpfulness and respectfulness on a five-point scale (A, B, C, D and F, like a class grade). The four grades given by each respondent, one each for professionalism, responsiveness, helpfulness and respectfulness, were averaged to create something akin to a grade-point average, much like the measure students worry about so often near the end of every semester.

The survey project was able to secure 973 completed responses to these four questions (many were incomplete, which is an issue worthy of consideration at another time, as is the question of whether some non-response bias exists, making the $n = 973$ less than random). The mean "GPA" for respondents was 2.784 (something like a C+), and the standard deviation was 1.023. We want to bracket off a 95% confidence interval for our approximation of the likely GPA for the whole community.

$$CI_{95} \approx \bar{x} \pm 1.96(\sigma_{\bar{x}})$$

So, we need the mean (2.784), the n (which is 973) and the standard deviation (which is 1.023). Let's first calculate the standard error. That's the standard deviation divided by the square root of the sample size. For this case, that means it is $1.023/\sqrt{973}$. That comes out to be about 0.0328, which is our standard error of the estimate. Multiply our z value for confidence, 1.96, times the standard error, 0.0328 and we get 0.0643 (rounded up).

We're almost there. The hard work is done. Our mean was 2.784, and our margin of error is 0.0643. Now we just add and then subtract:

$$CI_{95} \approx 2.784 \pm 0.0643$$

Or

$$CI_{95} = 2.7197 < 2.784 < 2.8483$$

We can read this by saying that we think the mean "GPA" for police service (for helpfulness, respect, professionalism and responsiveness) among all members in the community is 2.784 (again, about a C+) but we are 95% confident that it is between 2.7197 and 2.8483.

From *Z* to *t*

Let's shift gears for a moment. We've talked about the normal or z distribution quite a bit so far: we discussed z scores in Chapter 4, and we have talked about the z distribution in this chapter. We know that because of the knowability of the landscape that is z, and because of the Central Limit Theorem, z serves nicely as a proxy for the sampling distribution we won't ever actually have, which itself serves to help us infer a population parameter we won't ever actually be able to measure. It turns out that z isn't alone in this. We have made reference to other distributions already. There are many different kinds, although we will only put a few to use in this book. One of those is the **Student's *t*** distribution.

The use of any particular distribution as an able proxy for a sampling distribution depends on conditions of sample and question. For example, *z* serves admirably as a proxy for sampling distributions when measurements are plentiful. In leaner times, and when sample size figures heavily into our tests of inference (coming up soon), we might choose to use a different distribution. Student's *t* distribution is perhaps the most significant example.

Now, *t* and *z* are both symmetrical and bell-shaped, and one might be forgiven for mistaking one for the other at first glance, but unlike *z*, *t* varies according to sample size, or more particularly, it varies according to degrees of freedom.

Degrees of Freedom

The idea of **degrees of freedom** is relevant throughout much of what is to come, here and elsewhere. Many if not most statistical tests performed on a sample require some consideration for the degree to which scores might vary within constraints. We might think of degrees of freedom in terms of latitude, the space within which scores can "choose" to be whatever they want, or the *potential* for noise in the system. That latitude is limited, of course, once we know the sample statistics that interest us. OK, that might sound a little weird, but let's take a look at a rather practical example.

So, let's say we have a range of eight scores: 5, 8, 6, 7, 4, 7, 9, and 5. The sum for these scores is 51, and so the mean is 6.375. Let's say, though, that we had an unknown *ninth* score, and that we also knew that the mean for the *nine* scores was 6.4444. There's only one number that ninth score could be: it *has* to be 7. So, if we knew the statistic before we examined each of the scores individually, each of them could have been (mostly) anything, *save for the "last" one*, which was in a way "dictated" by the mean and by the rest of the scores. For example, this range of nine scores, 3, 8, 10, 2, 2, 7, 13, 4, and 9 also sum to 58 and have a mean of 6.4444. The same is true for these: 0, 6, 13, 2, 1, 5, 16, 4, and 11. Once we pin down the mean and eight of the nine scores, though, we can mathematically deduce the score for the last case. That last case, once we know the mean and the rest of the case scores, is *not* "free" to vary. Consequently, when we calculate a sample standard deviation, for example, we subtract one from the sample size because that last case is not free to vary – it is not accurately understood as part of the noise or latitude represented by the range of possible scores.

Student's t and Confidence Intervals

Student's *t* varies by degrees of freedom – the shape of the distribution actually changes as sample size varies. Specifically, when *n* is small, *t* is platykurtic. A quick peek at Appendix B will reveal that, confidence level being held constant (scanning down a column), degrees of freedom get higher (sample sizes get bigger), the critical value of *t* gets smaller (meaning a score need not differ as wildly from the mean to be notable). Some of this is a tad unclear, I imagine, as a few key concepts have yet to be developed. We will get to them in Chapter 8. The point for us here is that when *n* is modest, we can create a confidence interval using a *t* distribution if we so choose. Instead of using 1.96 for our confidence value corresponding to 95% confidence, we need to choose the *t* value that corresponds to the proper degrees of freedom.

Let's say we have a small sample, only $n = 11$. The mean of the sample is 13.2, with a standard deviation of 3.6. We will calculate a 95% confidence interval using *t*. The mechanics are the same, except we need to refer to Appendix B for the corresponding t value. The sample size was 11, and so we need degrees of freedom for $n-1$, or DF = 10. For a two-tailed test at 0.05 (or 95% confidence), the corresponding t value is 2.228. The rest we have done already. First the standard error of the estimate:

$$\sigma_{\bar{x}} = \frac{3.6}{\sqrt{11}}$$

$$\sigma_{\bar{x}} = \frac{3.6}{3.3166}$$

$$\sigma_{\bar{x}} = 1.0854$$

Then the confidence interval itself:

$$\text{CI}_{t95} \approx \bar{x} \pm t(\sigma_{\bar{x}})$$

$$\text{CI}_{t95} \approx 13.2 \pm 2.228(1.0854)$$

$$\text{CI}_{t95} \approx 13.2 \pm 2.4183.$$

Therefore, our t confidence interval for our estimate is:

$$\text{CI}_{95} = 10.7817 < 13.2 < 15.6183.$$

Summary

Inferential statistics can be understood to exist in three different kinds: point estimates, confidence intervals and hypothesis tests. In this chapter, we have discussed two of these kinds of inferences, and in so doing, set up some of the key reasoning for the third kind of inference.

The term "point estimate" refers to a sample statistic, like a sample mean, and its relationship to a population parameter. For us to be able to have confidence that the sample mean is a defensible estimate of the population value, it should be unbiased so that it is properly understood as a member in good standing of the sampling distribution for that population and parameter of interest.

Having a defensible point estimate, worthy though that may be, is not sufficient. We can use the available information to maximal effect by creating a confidence interval around that statistic, such that we can be confident that, all things being fair and representative, and that nothing is acting on the sampling process or values, the population parameter will lie somewhere within that interval.

We can use this same reasoning with other sample statistics. More, we can use other distributions, like t, when sample sizes are small.

Degrees of freedom are an important concept, relevant to many statistical tests.

Main Take-Aways

- Point estimates, like means of randomly drawn samples, provide us approximations of the thing we seek to know: μ, or the parameter of the relevant population. Under circumstances of a properly drawn sample, our humble mean, a descriptive statistic, becomes something more useful with regard to supporting inference.

- Fully recognizing that any point estimate is a guess, confidence intervals operate to give us a window within which we can be relatively confident the population parameter resides.

Key Terms (in order of appearance)

μ (Mu)	confidence intervals	student's t
asymptotic	margin of error	degrees of freedom

Chapter 8

The Reasoning Behind Hypothesis Testing

Most of the reasoning we need to support drawing meaningful conclusions about a population from a sample is already in hand at this point, but there are a few last – and critical – points that need to be addressed.

Right now, though, it is worthwhile to note that not all statistical applications involve the testing of hypotheses. We do not always assume random selection of a sample or rely on the Central Limit Theorem and assumptions about the sampling distribution, and we are not always anxious about the concepts we will cover in this chapter. Often, it is the case that we are content to perform statistical analyses merely to examine relationships within a dataset and not draw conclusions beyond that dataset to some wider population. Perhaps the data in hand are a convenience sample and generalizations are not fully warranted, or perhaps the data before us represent a full universe of possible cases and so no inference is needed. In such circumstances, we might perform statistical tests, even inferential statistical tests, but employ them rather differently. More is necessary before enlarging on this point, but it is useful for us to offer something like a caveat regarding the limits and scope of statistical inference and hypothesis testing here before delving into the subject in earnest.

What, then, is hypothesis testing? Well, it is the process of setting up a reasoned argument, often from some kind of theory in advance of performing any statistical analyses. And yes, the timeline matters greatly. Consider: for many years, I've played in rock-and-roll bands. Most of the time, shows occur at venues noted for their ribaldry, consumption of copious amounts of alcohol and, significantly for current purposes, the playing of billiards. Now, I cannot play billiards with any skill, really at all. One of the things I have learned from misspent years playing loud rock in honkey-tonk bars is that when cash is on the line, often the better pool players require each other to call their shots. That means they have to say what they plan on doing before they do it. "Four-ball in the corner pocket" is a thing that must come before the four-ball rolls across the felt and tumbles into the corner pocket. As crappy a player as I am, if I called my shots after I sunk them, you'd be foolish to give me credit for having understood the table sufficiently well to perform an act that altered that table in predicted ways.

And herein lies part of the key: a hypothesis is a statement derived from some understanding of the universe (or part thereof). That means we craft our theory, *and then* our research hypotheses *before* we perform our analyses. Otherwise, we may be engaging in what has been called "HARKing" (i.e., Hypothesizing After Results are Known) or engaging in a "fishing expedition." Let's be clear, though: there are times when we begin by doing research and then deriving insights from the results. Such an epistemic approach is called grounded theory (Glaser & Strauss, 1967/2012) and is a well-regarded and effective way

DOI: 10.4324/9781032638591-8

of knowing things. *It is in no way* my intention to imply that other epistemic approaches aside from a positivistic, deductive-then-inductive, hypothesis-first approach to science are of little value. That would be simply untrue. Indeed, part of the very reason for my writing this text is to frame more clearly what I take to be the relatively limited operable space for quantitative reasoning in the human sciences.

Notably, as much emphasis as I have placed on the importance of context and comparison in this text, comparisons are at the very heart of how many qualitative analytical strategies operate, including grounded theory in particular (Hallberg, 2009). Comparisons are common among many epistemic modalities, but the substantive richness of the comparisons possible in non-quantitative contexts tends to far outstrip the richness of comparisons possible with most statistical analyses.

Statistics, I have come to understand, do not represent the most fertile tool for understanding human harm and healing, but statistics are wonderfully useful in the spaces where they operate properly. My point here is to make clear that beginning from a hypothesis-first model, altering or adopting new hypotheses part-way through in order to make it appear as if one had "gotten it right" is reprehensibly bad science.

Concerning Hypotheses

In the most basic sense, a **hypothesis** is a statement of a relationship. It is a particular kind of statement and of a particular kind of relationship, though. We can envision a hypothesis as a statement at something of a crossroads between two axes. **The first axis is the theory-application axis** (see Figure 8.1). Hypotheses operate like corollary propositions that are subsequent to theory and seek to put ideas in the theory to the test of practical application in the real world. Often, we think about theory as being a thing that is highly, even hopelessly abstract, something of little if any practical value. Not so. Indeed, theory is *richly* useful, or else it does not deserve the distinction as theory. Not all theories are created equal, of course, but "theory" read generally is (or can be) the amassed and refined wisdom of the generations before ours. Certainly, such a thing is worth our time, don't you think? OK, I can hear you shrugging. In fairness to your disdain, particular social theories, and, in particular, some *criminological* theories, have not always held up their end of the bargain. We are not here to defend all theoretical endeavors merely because they exist. Indeed, I am partial to certain theories of and about justice (e.g., DeValve, 2015; DeValve et al., 2018), but I prefer those theories precisely because most others on offer have seemed to me rather disappointingly anemic and underpowered, but that's all a matter for another venue, another time. That particular theories fail is hardly a reason to discard the baby with the bathwater. Theories must sing for their dinner; if they aren't worthwhile, toss them aside. Mine too.

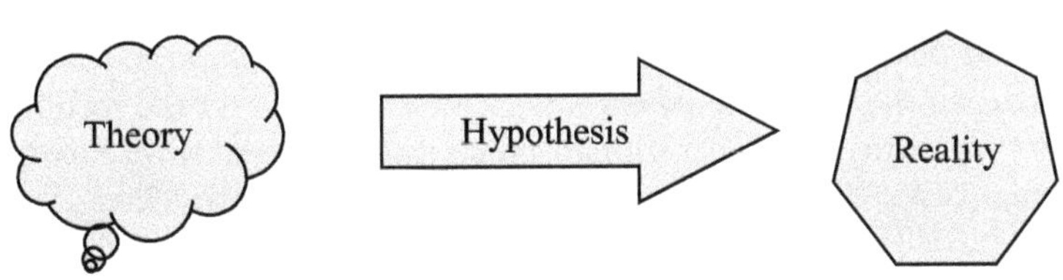

Figure 8.1 The First Axis: Hypotheses as Bridge between Theory and Reality.

But of course, that's precisely the point, isn't it? Theories must prove themselves useful; an idea as a map to the universe must be reasonably able to get us where we are going in that universe, or else it isn't a map to that universe. A theory might be analogous to a GPS device's software: it is good precisely and only when it helps us get from point A to point B. If the map is out of date, it doesn't work well and we get lost in strange and unfamiliar places. A good way to know if your GPS is up to the task of getting you home is, well, to head home with the GPS running, guiding, pointing. The map is the theory, and the empirical study is the actual driving. If the map is good, you get home. If it isn't, then you're stopping at a Waffle House and asking for directions. Pushing the analogy, it would be fair to say that criminologists have spent more than a little time, hat-in-hand, seeking guidance from short-order chefs in gray and yellow raiment. That, or they buck, and, like the archetypal stubborn father, drive in circles looking for landmarks while the family gets increasingly frustrated.

We should also point out again that theory needn't be formal criminological, social, psychological or other theory, written in a book or traced in a peer-reviewed article. Theory can be humble, it can be implied, subtle, and it can be all the more influential for being these.

There are a few aspects of theory use that can serve us well as we look to interpret statistical findings in terms of theoretical meaning. First, though, there are a few terms to get down. **Ontology** deals with thought about the nature of our universe and our place in it. **Etiology** is the study of cause, the degree to which a phenomenon is caused by other phenomena. When we talk about **criminoetiology**, we are speaking about the things that cause crime. It is notable, though, that we do not tend to talk about **hamartioetiology**, or the study of sin or flaw, and we are only beginning to discuss **zemioetiology**, or the study of the many causes of harm. **Epistemology** is the study of knowledge; how we know things or even what it means to know something. Not all theories are ontological; that is to say that not all theories intend to offer a map of all things or to explain the fabric of existence. Also, not all theories are etiological in nature, meaning that many theories do not necessarily seek to address causes of things. The thing is, though, if we are to assess the value of theory, be it ontological in nature, etiological in purpose, or both, our strategies of knowing things, our epistemic approaches to the world, need to be well in order before we begin.

Another thing to bear in mind about theories is that some theories, of both formal and informal kinds, seek usually to explain a broad class of events. They are concerned typically with treating a wide variety of phenomena. Darwinian explanations of biological adaptation seem to serve us well as an example. Darwin's overarching theory treats how adaptation occurs, but is less concerned why, say, a gerboa has no front limbs of which to speak. This grand-spectrum kind of theorizing might be thought of as **broad-class** or **top-down** theorizing. In contrast, some theories look at individual events and then craft the map they posit around that local context. We might call this **narrow-class** or **bottom-up theorizing.** This distinction between top-down and bottom-up theorizing is not world-shattering, but it can help us with interpretation of results in particular ways, ways that will become evident through the rest of this book.

The second axis describes the anatomy of a hypothesis. *A hypothesis is a statement of relationship between two or more constructs* (see Figure 8.2 and the formula that follows it). These constructs (often) are operationalized concepts derived from theory. If my theory (that justice and love are the same thing, and that individual heart-wellness is directly related to the quality of justice provided by institutions) has any merit, then a corollary argument is that police officers who practice some kind of contemplative practice on a regular basis and in a disciplined way will be less likely to be hijacked by fear, and thus engage in destructive,

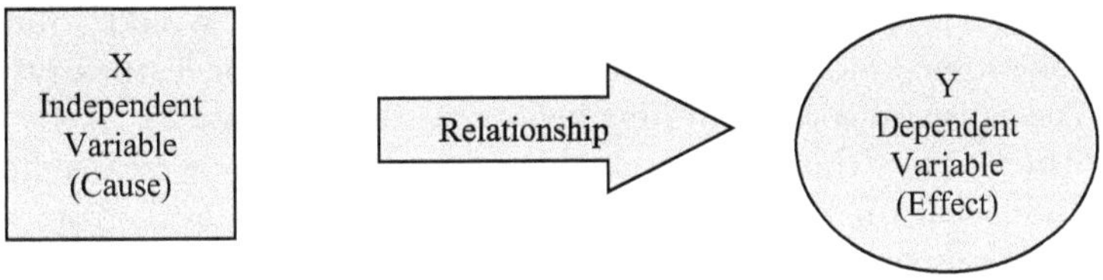

Figure 8.2 The Second Axis: Cause and Effect, Or Independent and Dependent Variables.

wisdom-abandoned behavior when confronted with difficult circumstances or people. If we can measure the amount and capacity of contemplative practice, and also measure acts of destructive policing, we might be able to show the theory's merit. One way to measure contemplative practice would be to count the number of hours in contemplative practice for each officer. We can measure acts of destructive policing by counting questionable uses of force and substantiated complaints. We might also use peer- or supervisor-assessments of officers' words and actions related to racism or other toxic behaviors. If the theory holds, we should see an **inverse relationship** between hours spent in contemplative practice and indices of destructive policing. An inverse relationship is in contrast to a **direct relationship**, where when *X* goes up, *Y* also goes up.

Remember before how we talked about populations? Recall that populations are best understood as semi-ethereal things, changeable and permeable. Because of their amorphousness, we need often to set out a sample frame – a particular iteration, clearly defined and empirically operable, of the population, so that we can study it. Well, we can think of a hypothesis as being analogous to that sample frame, whereas a theory, amorphous and ethereal, is analogous to a population. Indeed, *theories often posit things about populations directly*; one way to know if a theory has any value regarding the applicable population is to render from each an empirically operable version: a hypothesis (or hypotheses) from a theory, and a sample frame from a population.

Figure 8.2 shows the basic anatomy of a hypothesis: at least two terms (an *X* and a *Y*, *Y* being the effect and *X* being the cause) along with a statement of relationship. Another way to say the same thing presented in the formula:

$$y = f(x)$$

which can be read as saying *Y* varies as a function of *X*. If our theory has any merit, *X*, the independent variable (or a bunch of variables, together called a **variate**), is (or are) thought to have a particular impact on something – *Y* to be precise.

Just a Note about Causation

Although it is likely familiar, let us just take a moment to review the logic of causation. We can say with some measure of confidence that *Y* does indeed vary as a function of *X* only when three things are addressed:

1 **Correlation:** *X* and *Y* must have some form of demonstrable relationship. It might be direct, inverse, parabolic, sigmoid or some other thing, but *X* and *Y* should evidence a relationship in that they co-vary.

2 **Temporal Order:** If X is the independent variable, then it would strain credulity or perhaps invoke some kind of weird quantum entanglement for X, the cause, to follow Y, the effect. The cause must come in time before the thing caused.
3 **Nonspuriousness:** Now things get tricky here. Nonspuriousness means that there are no other rival influences on Y, no Z terms doing the influencing aside from X, and only as X works on Y. Simple enough, right? Well, that's really when the wheels tend to come off; how do you know you've measured all relevant variables that might work on Y? In a laboratory such a thing might be possible, but because we are talking about human sciences, there are both ethical and epistemic reasons why a laboratory just isn't so fantastically desirable. That means we might not ever be able to be certain that there is no Z term making things *appear* to be one way but in fact *are* some other way.

OK, I need to offer a brief comment. One of my mentors has written (e.g., Quinney, 2000) about "social causation;" he asserts, and I am certain that he's correct, that in human spaces, rules of causation are not nearly as straightforward as this three-part framework would have us believe. I say this not to complicate things, but merely to be forthright fully. It is important to learn this three-part framing for causation, but it is just as important to be able to transcend it later on. For current purposes, though, this formulation of causation is important to get down. Social causation, quantum entanglement and the rest are for another conversation later.

Scientific Reasoning, Statistical Decisions and Falsification

The challenge before us here is to figure out how we might use all that has come before to make decisions about the evidence we derive from our empirical/statistical musings. We have determined that if we can: (1) derive an unbiased sample from a population and (2) use the information in that sample with maximal efficiency, we can deduce from that sample, as a member of the proper sampling distribution from that population, both (A) an approximation of the population mean (that is, μ), and also (B) an estimate of the precision of that approximation (that is, the standard error of the estimate). We have also been able to derive estimates of error in terms of confidence intervals around our point estimates. Soon, though, we will find ourselves confronted with making that third kind of inference – hypothesis testing. And in fact, it is indeed hypothesis *testing*; we will want to be able to determine whether our hypothesis is supported by evidence or not, and we need some reasoning that provides some sound and consistent logic for the determination of what results actually mean.

In the current context (as this is in no small measure a book about statistical applications in criminology), that process of decision-making hinges on something we call a **statistical decision**; we must find a way that is clear and defensible to all audiences to frame the results of the statistical tests we perform, to decide whether evidence in hand supports a hypothesis, or whether it does not. The tests themselves do not provide such a thing. We might say that we need a bit of reasoning to help us make a statistical decision.

As no doubt you're aware by now, human sciences like criminology are not precise in the ways, say, chemistry or physics are precise. To pretend that they are is folly. Still, some of the same reasoning regarding how natural and physical sciences make decisions regarding findings serves ably in the human sciences. All scientists are humans, as it turns out, and often humans prefer to have their own mental maps of the world (e.g., their theories) be supported by evidence. Aside from professional and esteem benefits, it is just cool to get it right. Perhaps especially in this particular moment of decision, our human failings might come to bear on us; marginal results might be read favorably for our theories, firmly contrary results might be impugned unfairly, and supportive evidence might well be amplified in our glee.

In fairness, physical scientists right along with human scientists have suffered moments of weakness and proclaimed marginal results or rushed to ill-guided judgment. In order to be sure we made the right call, that we made the correct statistical decision, we need to operate from something of a particular view. This view helps us stave off our own biases, and perhaps just as importantly, it helps demonstrate that lack of bias to crucial audiences (don't forget: science is at its core a grand conversation). This view of things has a few components that we need to identify at the outset before we put it all together:

1 We do not prove things to be true. Conditions can change and so too the causal forces at work on *Y*; with new contexts (times, places, people), new spurious influences can arise, and worse, never be measured. Instead of proving something true, we demonstrate *falsehood.* Basically, this process of scientific inquiry assumes that (A) no relationship exists and that (B) evidence must be overwhelming that it *does in fact* exist before we (C) conclude that it *might* exist. We call this process **falsification.**
2 Only when it is shown conclusively that there may be some kind of relationship between *X* and *Y* do we "say Uncle" and conclude that there may not be no relationship, but that perhaps some relationship may exist. The hypothesis of no relationship that works as the default setting for science is called the **null hypothesis.** Only when evidence is sufficiently robust that the null hypothesis is unlikely (say, 1 time in 20 due to chance alone), we falsify, or reject, the null hypothesis and instead adopt the (or one of the) **alternative hypothesis** (or hypotheses).
3 We are anxious to avoid concluding that a finding, or even a set of findings are definitively and objectively true for all cases and settings. How could we after all? Findings, in order to be "true" must be repeatable consistently and under a variety of conditions. If findings vary by context outside of what theory predicts, then they should be understood to be suspect. In essence, then, we are supremely cautious about drawing conclusions, particularly conclusions that favor our theories.

Looking back at the example regarding contemplative practices and destructive policing, one might feel strongly that one's ideas might bear empirical fruit (and thus invite grant funding for programming and more projects, more insights, more students, more esteem, – OK, perhaps that's just a tad cynical). Were we to go in assuming our theories were right, we are very likely to find the evidence we seek (Gould, 1981/1996). Even the best of intentions might work against us in our scientific endeavors. So, what we do is to begin with the assumption that contemplative practices will have no impact, or perhaps might even make cops worse. *Only* if the evidence is *so overwhelming* that we would be foolish *not* to conclude that something was at work will we conclude that there *might* be a relationship. Skepticism, in the contemporary meaning of the term, is one of the first and most vital tools of the scientist.

In terms of symbols, then, we symbolize the null hypothesis: H_0. The alternative hypothesis, the hypothesis of interest is often given the symbol: H_A. We can write our hypotheses for the contemplative practices in policing study thus:

> H_0: Contemplative practices have either no impact or a negative impact on the presence of destructive policing.
> H_A: Contemplative practices decrease the likelihood of destructive policing.

Hypothesis Sets and Directionality

There are a variety of ways to arrange our null and alternative hypotheses; which arrangement we use is also a function of theory, of the nature of the question, and of your preferences regarding how you ask it. We might call these different arrangements "hypothesis sets."

We can think of there being two basic kinds of hypothesis sets: **nondirectional** and **directional**. Of course, both kinds of sets should be **mutually exclusive** (meaning that an outcome can't be both true and not true at the same time) and **exhaustive** (meaning that all possible outcomes are enumerated).

What I am calling *nondirectional hypothesis* sets are hypotheses that propose a relationship has a particular value, and that all other values are, in essence, presumed by the null hypothesis. As an example, recall from the last chapter how our police department did a citizen satisfaction survey. Let's say that the chief of police expected that the typical citizen would give the police department a B+ grade on the professionalism, responsiveness, respect and helpfulness index (let's call this the professionalism index GPA). Let's say that corresponds to a 3.2 GPA equivalent with our measure. Now, first, we need to do a brief intellectual backflip: the chief wants to be right here, and a B+ isn't a bad grade. So, in order to protect us all from our own biases toward what we hope might happen, we have to recall that the H_0 really comes first here. In terms of hypotheses, then, we would express that nondirectional hypothesis idea this way:

H_0: Professionalism Index GPA ≠ 3.2
H_A: Professionalism Index GPA = 3.2

In essence, then, if the professionalism index GPA derived from our sample is anything aside from 3.2 or thereabouts, we must accept the null hypothesis. This hypothesis we would call nondirectional (because it is really operable in *both* directions from the hypothesized value).

But hold the phone... surely the chief would be pleased if it were to be discovered that the professionalism index GPA is *actually higher* than 3.2, right? That would be reason for wholesome and ribald celebration. We might rethink our hypothesis above and make it directional, like this:

H_0: Professionalism Index GPA < 3.2
H_A: Professionalism Index GPA ≥ 3.2

In regular terms, then, our null hypothesis (our starting assumption) is that the professionalism index GPA is less than 3.2. The alternative hypothesis is that the professionalism index GPA is at least 3.2 – that it is 3.2 or higher. This is a *directional hypothesis*. We would be able to reject H_0 only when we have conclusive evidence that it is not the case, but even then, we are only adopting the alternative hypothesis in a very tentative manner; we might say that evidence indicates that H_A may not be false, *that it is not not true*.

So now the question arises: *just how much lower* than 3.2 is important? *Any* amount lower? That seems silly. How much lower must the score be to be *meaningful*? If you recall, the sample result for this measure was 2.78, which is lower than 3.2. Is it meaningfully lower? Should the chief be concerned? Should she be fired? Well, that's where the idea of statistical significance comes in.

Statistical Significance

Well, again, if you're seeking meaning, statistics isn't the place to find it. I hate to be the bearer of bad news, but meaning isn't what statistics does. Instead of meaningfulness, though, statistics can speak to likelihood, and there are more than a few times where something being unlikely due to chance can serve as a defensible proxy of meaningfulness (provided that we continue our search for meaningfulness). More to the point, *something being relatively unlikely invites the conclusion that perhaps something meaningful is occurring.* What that thing is, of course, is not the business of statistics. In fact, it is useful to think of inferential statistics as a set of formal logic processes, on either side of which rather less formal human processes occur. We build models based on our needs and interests, and we interpret results on the basis of what we put in and what it all might mean in a grander sense. **Make no mistake: statistical tests themselves do not decide things, and they do not simply output truth.**

So, if statistical techniques don't make our decisions, if they only speak to relative likelihood of outcomes, – whether something is likely or not – then it might make sense for us to have some kind of semi-bright-line standard that can be used and agreed-upon widely. And it turns out that we do have just such a standard. We call that standard **statistical significance.** When something is found to be sufficiently unlikely to be the result of chance alone, we deem it to be statistically significant (sometimes we say "nontrivial" just to impress our friends).

My grandmother worked hard to impart to me that nothing is ever certain. One hundred percent confidence in a finding is both an extraordinarily high bar to clear for most questions related to uncertainty, and also probably practically impossible to ascertain. Instead, for most social scientific applications, we have adopted a 95% confidence standard. Notably, the same 95% confidence standard operates in criminal courtrooms, and is operationalized there as certainty "beyond a reasonable doubt." It is not certainty beyond *any* doubt, after all.

When we talk about this standard of certainty of findings, we often use the shorthand p, or the probability of a thing happening due to chance alone. So, for a 95% confidence standard, we would say that the probability of a thing happening due only to chance would be 0.05. Thus, we refer to the $p = 0.05$ standard. This standard is a generally adopted (but not enforced or universally appropriate) rule for making statistical decisions. It must be noted, though, that $p = 0.05$ is not the only thing we consider when we make our statistical decisions regarding the null hypothesis (to wit, whether it should be rejected or not). Again, statistical techniques do not simply secrete wisdom; it is up to us to understand the processes that came before along with the relevant statistical process we used in order to make our statistical decision regarding the null hypothesis. The $p = 0.05$ standard can help us, but it is not the only consideration, and neither is it a hard-and-fast rule. Increasingly, researchers use the term "marginal significance" when findings have a p value in the neighborhood of 0.05 but are just slightly more likely, like $p = 0.063$.

Something of a warning is appropriate here, though. As Ziliak and McCloskey (2011) so vociferously observe, statistical significance speaks to *whether or not* a thing may exist, and not to the size or scale or function of the thing. Credibly we might ask whether or not a particular criminological theory has an explanatory shape on a crime/harm phenomenon, or we might ask whether or not a particular carceral program has an impact on recidivism, but questions related to whether these things are ethically defensible, authentically healing, maximally efficacious, economically sustainable, or questions in terms of *how these things substantively function in the world* ... are fully beyond the logic of statistical significance. Ziliak and McCloskey go so far as to describe significance testing as being a parlor trick elevated to the status of science. In fairness, they're not mistaken. It bears observing, though, that *whether-or-not* questions have been at the heart of what has had to be done criminologically during the 20th century. Let's not forget that although criminological pursuits date back centuries, the discipline as we know it today took a considerable portion of its current shape from events and people in the early-middle of the 20th century answering *whether-or-not questions*, which they had either inherited or developed to address pressing problems of the day. There was a lot of *whether-or-not-ing* to do even as recently as the 1980s. Much of the emergent shape of criminology is tied to the disappointing results from many of the *whether-or-not* pursuits that stand now like Moai, windswept and silent, speaking to no one.

Well, maybe not *no one*. The Moai of Easter Island meant something to someone, and the Moai can show us things relevant to us right now even though their sculptors have long vanished into the salty haze. There remain *whether-or-not* questions in criminology, and such questions will remain with us for some time, I imagine. Additionally, we should not take from these observations the conclusion that early criminologists were thinking poorly or were incompetent. There is much to learn from visited ground, even when that ground is largely barren (see, e.g., DeValve et al., 2018). Neither should we draw the conclusion criminology isn't "scientific" (it is, but it is also just as much a member of the humanities), that criminologists aren't doing their jobs, or that things are just too sophisticated for us to sort out. What "science" means might require us to think more broadly and more carefully than we have up to this point. No, statistics represents a vital and powerful way of knowing. It just can't show us everything ... or most things if we're completely honest. In the past, when positivist-scientific ways of knowing were the primary way of discerning any kind of truth (because truth was objective, and science succeeds – just look at all that science has done for us), it is only reasonable to build a scientific apparatus for criminological knowing. That it is only partly useful is no fault of the idea or of the architects. It *is* a problem if we learn to rely solely or primarily on statistical significance or over-quantified approaches to knowing about human harm and healing over ways that presuppose connection, nuance, meaning. Having done so in the past has represented loss, injustice and even death (Ziliak and McCloskey, 2011).

Figure 8.3 illustrates how all of this operates. Our sample is thought to be a legitimate member of the sampling distribution, and, thus, it contains values that approximate that sampling distribution (because, in theory, they contribute to it as a function of sample size). Our observed result from our statistical test can be held up to a probability distribution (in this case, we are using the normal distribution) because the Central Limit Theorem tells us that sampling distributions become normal as n increases. An observed score (expressed as relative likelihood in terms of where it falls along the x axis in Figure 8.3) has a corresponding p value (we might also remember that there's also a corresponding z score – an expression of the same thing in terms of standard deviations). Well, if we adopt the $p = 0.05$ standard for significance, we can also add something of a threshold here; if the score falls above that threshold (as it seems to do in Figure 8.3), it is falling in what we call the "rejection region." That means that a score that falls above that threshold invites us (but does not necessarily compel us) to reject the null hypothesis (H_0). If our observed result, complete with corresponding p value falls *below* that threshold, we are encouraged strongly to accept H_0 and conclude that the result is not significant, meaning that it does not vary from expected scores to a meaningful (read likely enough) degree.

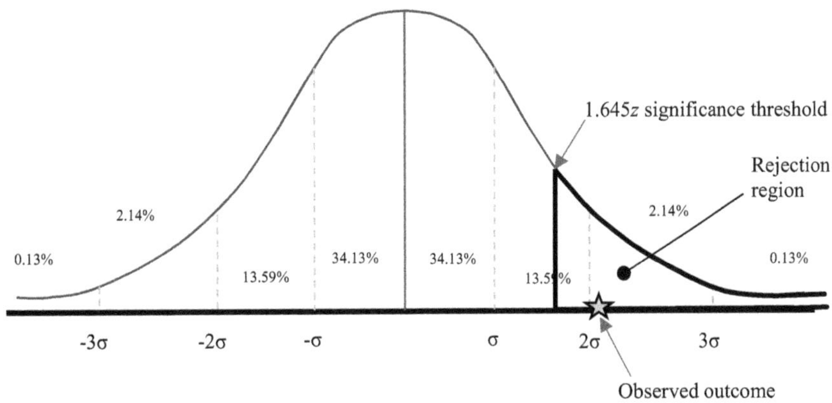

Figure 8.3 Statistical Significance and Rejection Regions.

Now what we see in Figure 8.3 is what we would call a **one-tailed hypothesis** (OK, and I should add that the proportions might be a little off. It looks like the area bounded in bold is a bit bigger than 5% of the area under the curve). It is directional, meaning that the rejection region is all on one side of the distribution. If you look at Appendix A, you'll see that the area under the probability curve corresponding to 1.64*z* is 0.4495. Well, 1.645 standard deviations above the mean marks out almost exactly 95% of the distribution. Below 1.645*z* lies 95% of all scores. That means a *z* score of 1.645 will happen by chance alone only one time in 20 (five times out of 100) due to chance alone. We might feel confident, then, concluding something else other than chance might be driving that outcome.

Recall from our conversation about confidence intervals how we used the value 1.96*z*? Well, 1.96 as a value has come back. It corresponds to 95% confidence for a **two-tailed hypothesis**. The 1.645*z* falls at 95% because only 5% of the distribution lies above it. If, however, we sought to ask whether a score is *different*, *either* higher *or* lower than expected, we would need to split that 5% rejection region and place half on either side of the distribution. Now, only 2.5% of the curve will lie beyond each rejection region threshold we have established, but as there are two now, the total rejection region turf is the same. That rejection region threshold would be ±1.96*z*. As we have seen already, of course, the threshold (or critical) values of *t* vary according to degrees of freedom.

There's a quick and easy way to determine the directional nature of any hypothesis. If the hypothesis uses the words "greater than," "less than" or the like, it is one-tailed. If it uses the phrase "different from" or something similar to it, it is two-tailed.

Our concerns about confidently rejecting the null hypothesis ($p = 0.05$), important though they are, can overshadow another kind of error.

Types of Errors and the Error Grid

It turns out, in fact, that there are *two* general kinds of errors that should concern us regarding our statistical decisions. We will do well to be concerned about the probability of each kind of error – individually, and also considered together.

The first kind of error is the rejection of H_0 when it is true; when we in error *reject* the null hypothesis. We call this kind of error **Type I error**. The probability of Type I error is

		Actual Reality (what we should have decided)	
		H0 is actually true	**H0** is actually false
Statistical Decision	Accept **H0**	Success!!	Type II Error (β)
	Reject **H0**	Type I Error (p, or α)	Success!!

Figure 8.4 The Error Grid.

p (sometimes we also refer to that probability as α, the Greek letter alpha). In contrast, **Type II error** occurs when we mistakenly *accept* the null hypothesis when we should reject it in favor of the alternative hypothesis. We use the symbol β (the Greek letter beta) to represent Type II error (Figure 8.4).

Again, we are perpetually concerned about the relative likelihood of each of the kinds of errors.

For Type I error (α), the probability is the same as the p value we chose for our threshold (or critical) value. If we opted for $p = 0.05$ for our two-tailed hypothesis, our critical value of z will be 1.96 (corresponding to 0.4750 in Appendix A, or 47.5% of the distribution lying between the mean and that point).

Things get a tad trickier with calculating the probability of Type II error (β), however. To know the probability of Type II error, we need to have (or approximate) several measures: (1) the p value chosen; (2) the sample size; (3) the standard error; and (4) the difference between the actual parameter and the observed statistic (which we will call the **effect size**, something we will discuss in detail later in Chapter 9). Clearly, we aren't going to have the last two. We can estimate the standard error, of course, but if we had the population parameter, we wouldn't need to do any of this.

Statistical Power

If β is the probability of not finding an existing difference, then $1 - \beta$ is the probability of finding one if in fact it exists. This probability, the probability of finding something if it is there, is called **statistical power**.

The emphasis here will be the presentation of statistical power in conceptual terms, followed by a treatment of an approach, complete with formula that has relatively broad utility. First, though, let's think about power this way:

Looking at Figure 8.5, we can see two curves. The one on the left, the one in red, represents the null hypothesis (H_0) sampling distribution, and the curve on the right, looking sharp in black, is the sampling distribution for the alternative hypothesis (H_A). In terms, this means that the curve on the left is the sampling distribution in the circumstance that there is no effect, or no difference between what we think might be true and chance. If H_0 is true, then setting our critical value for z where it is means that we will accept H_0 95% of the time, meaning that we will conclude that there is a difference when there actually isn't one, only with one sample in 20. We get to determine where that line falls. In this case, we can tell it is a one-tailed test because there is only one rejection region (that's the chunk of the curve lined out in red. In fact, our α value ($p = 0.05$) and our sample size are really the only things we can practically control regarding power. Increasing α (which would mean sliding that vertical middle line to the left in Figure 8.5) might help increase power (meaning it would increase the likelihood of finding a difference if one actually exists), but by definition, it would also increase the probability of Type I error (that shaded rejection region would get bigger, which we don't want).

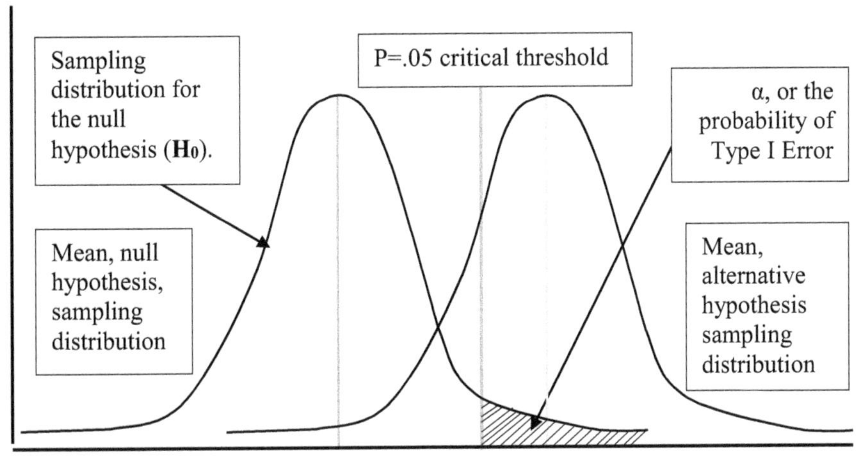

Figure 8.5 Statistical Power and α.

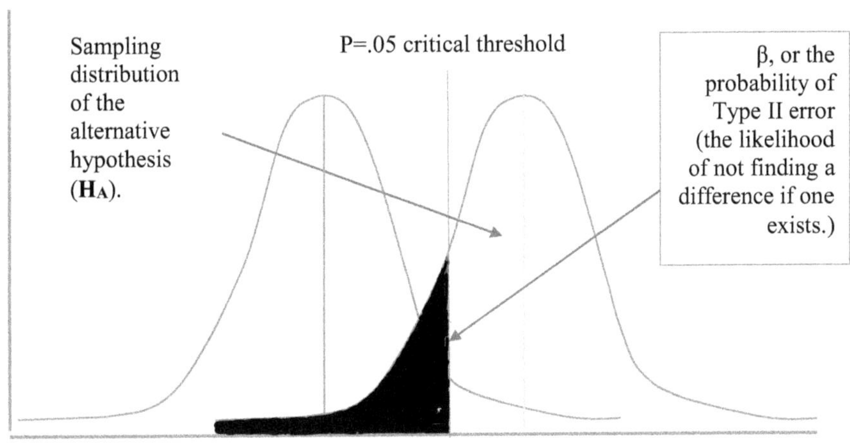

Figure 8.6 Statistical Power and β.

Figure 8.6 draws attention to the space that corresponds to Type II error, or the likelihood of missing an important (meaning unlikely) difference. A difference between the two sampling distributions exists, the left one is nontrivially different from the right one, but we missed it. If we moved the middle line to the left, we would decrease the probability of Type II error, but also increase the risk of Type I error.

The shaded area in Figure 8.7 represents the probability of successfully identifying a difference between $\mathbf{H}_0$ and $\mathbf{H}_A$ if it actually exists. We can see in Figure 8.8 the two means of the two distributions, and the difference between these two means is what we call the effect size.

Statistical power is an important if sometimes misused concept. For example, I have been asked more than a few times about identifying the necessary sample size for a certain power level (say, for example, 0.8). Unfortunately, such a thing is notoriously difficult to

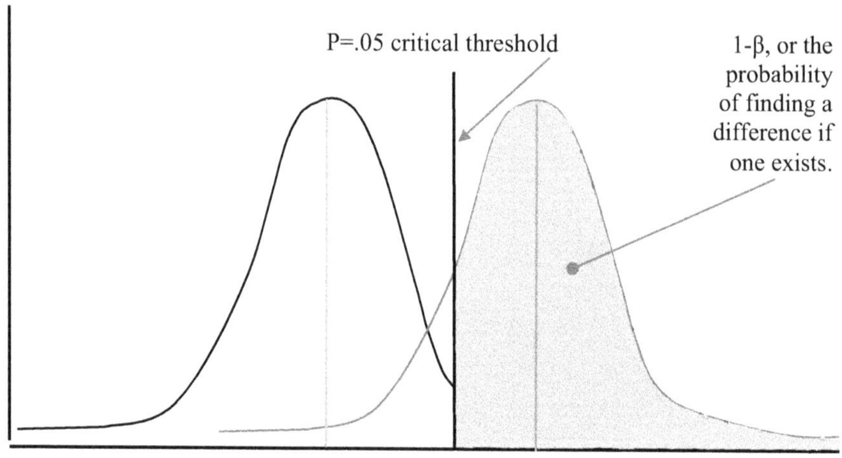

Figure 8.7 Statistical Power.

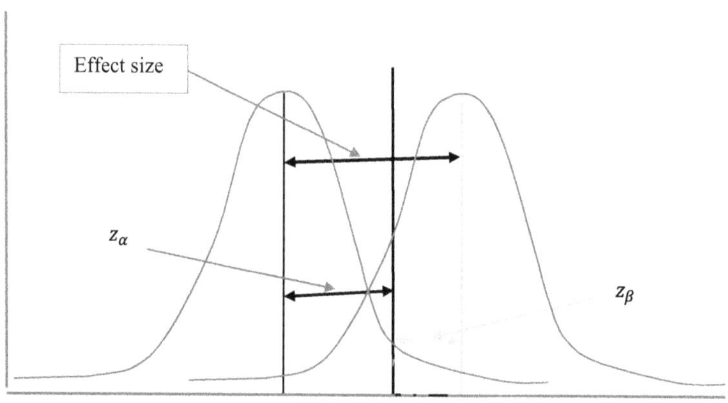

Figure 8.8 Statistical Power and Effect Size.

discern in circumstances where sampling is not unbiased, and it is almost pointless in fact when generalization is not sought. Contemplating statistical power can be a useful exercise when considering how to frame results, but outside true random designs, power is rather less meaningful than one might hope. As we have indicated already, in practical terms, the only thing we can impact directly as researchers is sample size (and of course, we can dial back our *p* value, but we do so, we've seen, at our peril). And again, we can't know the alternative hypothesis mean and sampling distribution. If we did, we would not need to have this conversation. We can approximate power if we know something reliable about the $\mathbf{H}_A$ sampling distribution, but perhaps it is more useful to talk about the sample size we need for deriving a desired capacity to find a difference, say, between two groups or between a sample outcome and a hypothesized value.

We can estimate the sample size we need for a desired power (Kachigan, 1986) using:

$$n = \frac{\left(z_\alpha - z_\beta\right)^2 \sigma^2}{\left(\mu_1 - \mu_2\right)^2}$$

where n, of course, is the sample size; z_α represents the z score that corresponds to Type I error (meaning the p value we choose – one-tailed or two-tailed); z_β corresponds to the z score equivalent of the space between z_α and the hypothesized mean of the alterative hypothesis sampling distribution (we might very well be guessing at this mean) (see Figure 8.8). The last term in the numerator is the sample variance. The squared difference between the two means (μ_2 again might well be hypothesized) comprises the denominator.

It deserves noting that different statistical tests have their own more precise ways for determining sample sizes, and Cohen (1988) presents both the formulae used and also some rather convenient tables for the determination of proper sample sizes given other terms in the setting. Kachigan's formula above, though, represents a generalized tool when tables aren't handy, and it serves well for most purposes.

Let us consider an example. Our chief asserted that the professionalism index GPA for the police department was a B+, or around a 3.2. The actual score, the one we measured in our sample, was 2.784 (sample $\sigma = 1.0232$). We can determine just how many randomly drawn cases we would need to have to find a significant difference if it exists in fact, to see if the chief is right or not.

First, we need z scores for α and β. Because the question is one-tailed (clearly 3.2 is larger than 2.784, and our H_A is that the GPA is equal to or greater than 3.2), we will use $1.645z$, placing our whole 5% on one side of the curve. What we don't know is whether the sample we have, with a mean of 2.784 is meaningfully different (that is, sufficiently unlikely to be the result of chance) from 3.2.

As we have indicated, we will need to squint a little to make this work. For starters, recall that we need values for z_α and z_β. Because it is a one-tailed test, we will use $z_\alpha = 1.645$. z_β is determined by our desired power. We are going to strive for a power of 0.75, or a 75% likelihood of finding a difference if one exists. Often, 0.8 is the standard for power, but I am choosing a slightly lower power level for this example. Recall that power can come at the expense of the risk to Type I error; the effect size in this case is not terribly large. Referring to Appendix A, approximately 25% of a z distribution falls between a mean and $0.68z$, so we will use –0.68 to give us a power of about 0.75 (75%, or we will correctly discern a difference if it exists 15 samples out of 20). What we will get, then, is:

$$n = \frac{\left(1.645 - (-0.68)\right)^2 \times 1.0232^2}{(2.784 - 3.2)^2}$$

$$n = \frac{(2.325)^2 \times 1.047}{0.1731}$$

$$n = \frac{5.406 \times 1.047}{0.1731}$$

$$n = \frac{7.706}{0.1731}$$

$$n = 43.94$$

We can be 75% sure (again, 7.5 times out of ten) that a sample of at least 44 cases (*randomly drawn*) will produce a correct determination regarding the null hypothesis. The n for the study was well more than that ($n = 973$ usable cases for that variable). As it turns out, the 3.2 to which the chief alluded is only 0.407σ above the mean, thus it is plausible that the outcome he celebrated is well within possible parameters. Don't forget:

the standard deviation for the sample was 1.0232, so a score of 3.2 is not all that far away from the mean. Did the chief fib? Maybe. Was his answer statistically flawed? We shall see in Chapter 9.

Summary

Statistical analyses do not generate or secrete wisdom. In order for statistical analyses to be able to work ably to support human decision-making, several key ideas are needed. In the narrow and rather particular setting involving positivistic quantitative human sciences, we benefit if we make certain things clear at the outset. In essence, we need to set up the reasoning framework within which statistical analyses and their results can be utilized.

Ontology, the study of the nature of existence, is not a thing typically touched in detail by statistical analyses, although it would be unfair to say that statistical analyses do not reveal things of ontological significance. Questions about cause, or etiology, are closer to the wheelhouse of statistical capacity, although we do well to heed Ziliak and McCloskey's warnings. Statistical techniques and reasoning operate as one of several epistemological frameworks within which we might undertake scientific inquiry.

First, of course, we begin from theory; whether in the body of formalized ideas or of less formal program theory, extant wisdom provides a sound beginning point. From a theory, though, we must make a practicable hypothesis (or hypotheses); likewise, from a population we must also make a practicable sample frame. From that sample frame, a random (that is, EPSEM) sample of sufficient size is drawn. From that sample, we can derive both point estimates and a sense of the precision of those estimates.

But we can go further still. We can utilize the point estimates and the standard error of the estimate to infer things about the relative likelihood of outcomes indicated in our sample as being reflected in the population from which that sample came. We must be cautious, however, as we may remain tethered to our biases and proclivities. To shield ourselves from our own shortcomings, we can arrange our hypotheses and our statistical testing in such a way as to default to a non-finding, or null hypothesis; only when we can discern with precision that the null state is highly unlikely do we permit ourselves to conclude that the null state may not be an accurate portrayal of things in the population. Findings of high-unlikelihood, or statistical significance, permit us (but do not bind us categorically) to reject a null hypothesis. Falsely rejecting a null hypothesis, or Type I error has been the chief concern, but the probability of Type I error exists in relation to the probability of falsely accepting the same null hypothesis, an error known as Type II error. Our ability to be confident in our statistical decision (which is probably better understood to be a human decision informed by statistics), is in part a function of the size of apparent effects (or effect size) and the statistical power of a test in context (the ability to discern an effect if one exists).

Main Take-Aways

- Positivist, empirical, quantitative epistemic strategies are one among many useful ways of knowing things that matter for human actualization. Such strategies are not always the best way of addressing pressing human questions, although they work ably in the contexts in which they are a proper fit.
- Statistical techniques need to be nested in a reasoning system for them to be able to support the decision-making process.

- That reasoning system currently used is set up in such a way as to shield ourselves... from ourselves.
- Once all the reasoning is in place around the statistical tests themselves, though, statistical techniques can reveal things about a population on the basis of information in a sample – information beyond just approximations of population parameters. We can discern relationships, differences, and test whether such differences are likely to exist in the parent population.
- Historically, Type I error has been the primary concern of scientists, although in the context of applied human sciences, like criminal justice, criminology and zemiology, Type II error is as pressing a concern, and at times should be the chief worry.

Key Terms (in order of appearance)

hypothesis
ontology
etiology
criminoetiology
hamartioetiology
zemioetiology
criminoetiology
epistemology
broad-class theorizing
narrow-class theorizing
inverse relationship
direct relationship
variate
correlation
temporal order
nonspuriousness
statistical decision
falsification
null hypothesis (H_0)
alternative hypothesis (HA)
nondirectional hypothesis
directional hypothesis
mutually exclusive
exhaustive
statistical significance
one-tailed hypothesis
two-tailed hypothesis
Type I error
Type II error
effect size
statistical power

Chapter 9

Bivariate Hypothesis Testing

t Tests

We have reached something of a turning point in our time together. We have been contemplating the reasoning that grants us the ability to draw inferences about a population based on efficient use of results from an unbiased sample from that population. What comes next for us is to discuss the statistical tests themselves. Almost all the statistical tests that follow involve some kind of signal-to-noise comparison. More, the reasoning at work with *t* tests is foundational to a significant portion of Ordinary Least Squares bivariate and multivariate statistical techniques, particularly those which we will discuss here.

In this chapter, we will be focused on bivariate comparisons, particularly where the dependent variable is a continuous or metric one, and the independent variable is binary (meaning that it has two discrete values). We will discuss three tests, all of which are *t* tests. We call them *t* tests because we use the *t* distribution for interpretation purposes. They reference a distribution called Student's *t* distribution because William Sealy Gossett, the inventor of the techniques, was forced to publish his ideas about the test and the distribution anonymously because his employer (Guinness beer) required it (e.g., Ziliak & McCloskey, 2011); the article that introduces the test was written by a "Student" – Gossett himself. All three tests work in roughly the same way, comparing a sample mean to something, either a hypothesized value or another sample mean, and considering that difference in terms of an expression of standard error. The result of the calculations we do will be a value, a number, which we can compare to a threshold or critical value at the correct degrees of freedom to determine its relative likelihood of arising. These three versions of the *t* test are made for application to different kinds of questions.

The level of measurement of variables involved in our tests is key, of course, but perhaps most obviously so for the dependent variables we feed them. The dependent variable *must* be metric because the variable must be able to produce meaningful means. The independent variable will be binary, as we have indicated already.

The three tests all follow the basic form:

$$t = \frac{\mu_1 - \mu_2}{\text{se}}$$

where the numerator represents a comparison of two means (or a mean and a hypothesized value), and the denominator contains a measure of standard error (that's what "se" stands for). It is, quite literally, a signal (difference between two means) and noise (standard error) ratio. The signal is the difference between either observed and expected scores (say, a mean and a hypothesized value) or two group means, and the noise is the standard error, or the amount of rattle-and-hum present in the two variables. As indicated already, the result of

DOI: 10.4324/9781032638591-9

the equation is a number, a difference expressed in terms of standard errors from a hypothetical sampling distribution. Once we consider the final bit of information – degrees of freedom – we can look up the critical value for *p*, and then compare the observed result (the number we calculated) to the critical value from the *t* table to derive an outcome that will guide us in making a statistical decision. **If the observed value, the number we calculated, is bigger than the threshold critical value of t that we've chosen based on our p value and degrees of freedom, then we can be on solid ground in concluding that something other than chance is at work, and that the difference might be the result of something aside from random sampling.** This process, the derivation of a number (often but not always a signal-to-noise ratio), which is compared with a probability curve that stands in for a sampling distribution (as context), will be with us through most of what follows.

Single-Sample t Test

The most basic version of the *t* test is the single-sample *t*:

$$t = \frac{\bar{x} - \mu}{\sigma_{\bar{x}}}$$

where $\bar{x}$ is the sample mean, μ is the hypothesized population parameter and $\sigma_{\bar{x}}$ is the standard error of the estimate (which is, again, our estimate of the standard deviation of the sampling distribution). It is called a single-sample *t* because, well, it uses only a single sample. Of course, because it involves a single sample, this particular *t* test is a univariate, not bivariate, test. It is presented with its kin, however, due to its clear resemblance to family, and because it illustrates the mechanism of the *t* test rather admirably.

We use single-sample *t* in circumstances where we have a single-sample mean of interest, a hypothesized value for the population parameter, and the standard error estimate derived from the sample we have. In essence, then, we use single-sample *t* to test the likelihood that our idea of the value of μ is appreciably different from what our sample indicates it might actually be by virtue of what the sample mean ($\bar{x}$) and standard error ($\sigma_{\bar{x}}$) together indicate it might be. Let's consider an example.

Consider an example: for the range of scores (6, 7, 4, 8, 6, 7, 9, 5, 7, 4, 0, 6, 4, 8, 9, 9, and 4), what is the probability that it is an unbiased sample from a population where the parameter is 3.0? Why 3.0? Well, that would come from theory ultimately, and would be specified via some formal hypothesis.

OK, so first we need that formal hypotheses. Let's frame our question in terms of a one-tailed test, so we will need to know the mean for the sample. The sum of the 17 scores is 103. 103 divided by 17 is 6.059. The standard deviation of these scores is 2.384. The null hypothesis is that there is no difference between the sample mean and the hypothetical population parameter where $\mu = 3$, and that any differences we do see can credibly be attributed to sampling error – that rattle-and-hum we saw at work back when we flipped coins. The implication is that the sample comes from that larger population, or that there is no meaningful difference between the sample and the population. To say the same thing, we could say that the sample we have is a proper random sample from the sampling distribution of the population.

Well, we know the mean is bigger than 3, so the alternative hypothesis, the conclusion to which we must be sold by evidence, is that the sample is not a member of the sampling distribution where the population parameter mean is 3.

Formally, then, we would write:

H_0: 6.059 ≤ 3
H_A: 6.059 > 3.

It sounds nuts to assert that 6 is equal to 3. Fair enough. And I'm not asserting *that.* Instead, we are asking how likely it is that we could have a random, representative sample with a mean of 6.059 that came from a population where the population mean (parameter) was 3. If the difference between the sample mean and the population parameter is sufficiently large, *given the amount of variance in the sampling distribution*, then we can conclude that something other than chance is at work in the derivation of the sample we have. It is fully possible to have a sample mean that is quite different from the population parameter if, say, the sample size is small or if the variance in the population is large. Thus, this is in no way an unreasonable question.

Again, our formula is:

$$t = \frac{\bar{x} - \mu}{\sigma_{\bar{x}}}$$

Recall that $\sigma_{\bar{x}}$, the standard error of the estimate, is calculated:

$$\sigma_{\bar{x}} = \frac{\sigma}{\sqrt{n}}$$

So, then, for our circumstances in this example,

$$\sigma_{\bar{x}} = \frac{2.384}{\sqrt{17}}$$

or

$$\sigma_{\bar{x}} = \frac{2.384}{4.123}.$$

Which means our standard error of the estimate is 0.578. Let's finish our single-sample *t* test.

$$t = \frac{6.059 - 3}{0.578}.$$

We can read this in terms of the difference between the discovered mean and the hypothesized value of the population mean, divided by the standard error, or the average amount of natural variation in the sampling distribution.

Plugged in, we have:

$$t = \frac{3.059}{0.578}.$$

So, our observed t value (often noted as t_{obs}) is 5.29. What now? Well, first, let's return to the reasoning of the process so far. That number, the 5.29, is understandable in terms of a signal-to-noise ratio. For every one unit of noise (meaning for every unit of standard error – for each approximate standard deviation of the sampling distribution), there are five units of signal. The difference between the observed mean and the hypothesized mean was pretty big relative to the approximated amount of variation. The big question, though, is whether or not the difference is big *enough* (really, if the difference we see is big enough to be vanishingly *unlikely*) in the context of approximated standard error to reject the null hypothesis. What we need, then, is a critical threshold score for t, what we call t_{crit}.

Recall that the t distribution changes its shape as a function of degrees of freedom, so we need to know n. Actually, degrees of freedom for a single-sample t is $n-1$ (again, because once the sample statistics are known, all but one score could take any value). To get our t_{crit} value, we need to have our degrees of freedom, our test directionality (one-tailed or two-tailed), we need our p value (the default is $p=0.05$), and of course we need our t table. We have our degrees of freedom ($17-1=16$), our directionality (one-tailed), the p value (0.05) and the t table is Appendix B. If we find the one-tailed column at the top, and the column that corresponds to 0.05, and then scroll down to the row for 16 degrees of freedom, we see that the t_{crit} is 1.746. That means that about one random sample of a size $n=17$ out of 20 such samples from a sampling distribution from a population where μ is 3 will produce a mean that is 1.746 standard errors above (or below, but we're just concerned with whether the mean is larger than) the population parameter. In our case, the t_{obs} we found was 5.29 and 5.29 falls well outside the rejection region set by the 1.746 t_{crit} value. Because t_{obs} is greater than t_{crit}, we reject the null hypothesis (that the population parameter is 3.0 or smaller) in favor of the alternative hypothesis (that the population parameter is greater than the hypothesized value). That doesn't mean we've proven that the population parameter is greater than 3.0, though. We can only say with certainty that based on the sample, the population parameter is unlikely to be 3.0 or lower. As we've seen, repeated samples can give rise to rather differing outcomes.

Here's another example:

The last chapter ended with something of a cliffhanger. The mean professionalism index GPA for the sample was 2.784, with a standard deviation of 1.023 ($n=973$). The chief had asserted publicly that the department had earned a B+. We can figure that a B+ is roughly a 3.2 GPA. In essence, then, we want to know whether a population where the real mean (μ) is 3.2 could produce a sampling distribution where an unbiased sample could have a mean of 2.784 is defensibly likely.

The reasoning process for answering a question like this begins with the null hypothesis, and we will use:

H_0: Professionalism Index GPA >= 3.2

In this case, the null hypothesis is that the actual professional index GPA is *equal to or greater than* 3.2. Now, recall that the null and alternative hypotheses need to be both mutually exclusive and exhaustive. That means, then, that the alternative hypothesis must be every other possibility, or:

H_A: Professionalism Index GPA < 3.2

The question we are asking is whether the sample derived is nontrivially less than 3.2, so we need to arrange our test such that we have to climb the probability hill to arrive at a conclusion. In this example, the hypotheses are arranged in a *one-tailed* fashion: we are asking whether or not the empirical results from the sample are significantly lower than our hypothesized value.

Well, as we have seen, to do the single-sample *t*, we need the sample mean, a hypothesized parameter, and a standard error of the estimate, and then we need to know degrees of freedom, and finally a *t* table.

Our standard error of the estimate is calculated thus:

$$\sigma_{\bar{x}} = \frac{\sigma}{\sqrt{n}}$$

$$\sigma_{\bar{x}} = \frac{1.023}{31.193}$$

$$\sigma_{\bar{x}} = 0.033.$$

All of the ingredients are now in hand (again, Appendix B is our *t* table), so let's begin. Here again is our formula:

$$t = \frac{\bar{x} - \mu}{\sigma_{\bar{x}}}$$

$$t = \frac{2.784 - 3.2}{0.033}$$

$$t = \frac{-0.416}{0.033}$$

$$t = -12.606$$

So what? Rather, now what? What does this number mean? Again, this number is a signal-to-noise ratio, the difference we found between the mean and the hypothesized value expressed in terms of standard errors. 2.784 is 12.6 standard errors (not standard deviations, mind you) below the hypothesized value of 3.2. Now we need to know how likely such a thing is to happen due to chance alone.

First, though, we need to find out how many degrees of freedom we had. In this case, because n was so large, t operates almost identically to z (that happens around an n of about 120, as you can see in the bottom row of Appendix B). For a one-tailed test, the critical threshold for 95% likelihood ($p = 0.05$) is 1.645, just as it is with a normal z distribution. Well, −12.6 is considerably smaller than −1.645. Because the observed t (−12.6) is smaller than the t critical score (−1.645), we can reject the null hypothesis, and conclude that the sample mean we have is (highly) unlikely to have come from a population where the parameter was actually 3.2. Maybe the chief needs to cool her jets.

The Snacky Statistician: Single-Sample *t* Test Recipe Card

Basic Ingredients

One dependent variable, continuous (metric) level of measure

One hypothesized population parameter

One hypothesis (one-tailed or two-tailed, depending on preference)

Directions
With your dependent variable, derive a mean and standard deviation.
Then create a standard error for that same variable.
Combine wet and dry ingredients according to the formula.
Compare result to t distribution at n-1 degrees of freedom.
Derive statistical decision.
Serve immediately.

Independent-Samples *t* Test

What if instead of a sample mean and a hypothesized value, we want to know if there is a meaningful difference between *two* sample means? We can use the very same logic as we did for the single-sample t:

$$t = \frac{\bar{x}_1 - \bar{x}_2}{\sigma_{\bar{x}\text{p}}}$$

except that we will need to account for two measures of variability instead of only one, as before. Instead of a single standard error, then, we will pool, or average standard error ($\sigma_{\bar{x}\text{p}}$).

In Figure 9.1, we can see two distributions, each with a mean, and we can see each distribution has variance. These two distributions have about the same amount of variance, and indeed that similarity matters. Because we are averaging standard errors (actually, it is a weighted average akin to what we discussed in Chapter 3), if the two distributions are wildly different, t test doesn't work as well.

Now it turns out there are two ways to pool standard errors for independent-samples t. We will present them both, but we will use only one. The standard practice for pooling standard error is:

$$\sigma_{\bar{x}\text{p}} = \sqrt{\frac{\sigma_\text{p}^2}{n_1} + \frac{\sigma_\text{p}^2}{n_2}}$$

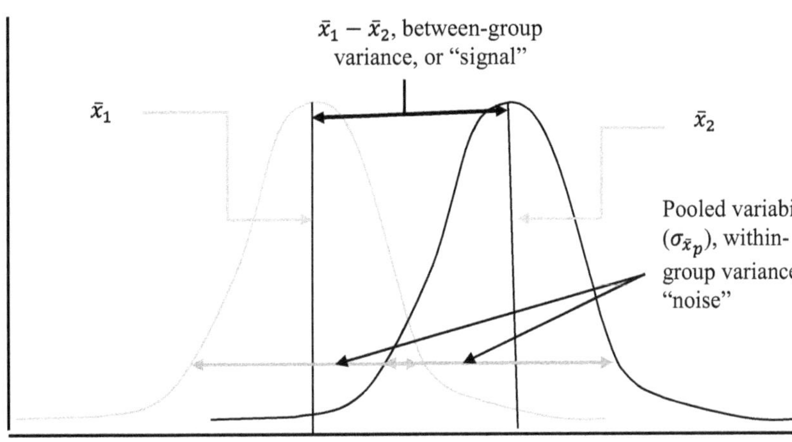

Figure 9.1 Independent-Samples *t* Test.

where σ_p^2 is:

$$\sigma_p^2 = \frac{(n_1 - 1)\sigma_1^2 + (n_2 - 1)\sigma_2^2}{(n_1 - 1) + (n_2 - 1)}$$

Here the σ^2 expressions are the variances for each sample, and n refers to each sample size. A more conservative and considerably less involved pooling of standard error (used specifically when equality of variance cannot be assumed) looks like this:

$$\sigma_{\bar{x}p} = \sqrt{\frac{\sigma_1^2}{n_1} + \frac{\sigma_2^2}{n_2}}$$

This expression, $\sigma_{\bar{x}p}$, is a pooled, or averaged, within-group variance for the two sample groups. Again, the σ^2 expressions here are the variances for each sample. In instances where equality of variance cannot be assumed, we can use this standard error. It is rather clearer and more straightforward mathematically while also being somewhat more conservative, so it will be the formula we use for hand-calculating independent-samples *t*.

The difference between the two standard errors is where the pooling, or averaging, happens. Notice how in the equation just above, each sample variance (itself a sum, recall, of differences from a mean) is divided by a sample size, just like the way a regular mean is derived. For the first version, because variance equality is assumed, we begin by smooshing the two variances together into a single term, used twice in the fractions under the radical. For the $\sigma_{\bar{x}p}$ we will use, the one just above, the two sample variances themselves are used directly in the calculation of standard error.

One last point to be made here is the calculation of degrees of freedom. Because we have two samples kicking about, we need to account for freedom in both variables. Therefore, the degrees of freedom value is calculated: $(n_1 + n_2) - 2$.

Let's try one, shall we? First, let's refresh the formula fully assembled (as we will be using it). This is what it looks like:

$$t = \frac{\bar{x}_1 - \bar{x}_2}{\sqrt{\frac{\sigma_1^2}{n_1} + \frac{\sigma_2^2}{n_2}}}$$

Let's say there has been a question about improving the wellbeing of police officers. It turns out that a considerable portion of the officers smoke. HR has records regarding the number of sick days used by all officers. We might use "number of sick days" as a proxy measure for wellness, and "smoke/don't smoke" as a measure of smoking activity. In essence, we

Smokers ($\bar{x}_1$):	*Nonsmokers* ($\bar{x}_2$):
Mean sick days per year: 3.74 Standard deviation = 0.98 $n = 40$	Mean sick days per year: 1.12 Standard deviation = 0.77 $n = 65$

are determining the likelihood that these two groups represent unbiased samples from the sampling distribution of the same population. Let's say our agency data look like this:

OK – now for the hypotheses. We can see that the mean sick days per year for the smokers' group was greater than the nonsmokers' group, so we can ask if the smokers' group uses significantly *more* sick days than does the nonsmokers' group. That makes our test a one-tailed test. We will use the usual $p=0.05$ standard as well.

> H_0: smokers' mean sick days **is less than or about the same as** nonsmokers' mean sick days
> H_A: smokers' mean sick days **is more than** nonsmokers' mean sick days

Or

$$H_0: \bar{x}_1 \leq \bar{x}_2$$

$$H_A: \bar{x}_1 > \bar{x}_2$$

Again, we think smoking officers need more sick time, but to determine whether they need *significantly* more sick time, we've arranged things just like a criminal trial – innocent (null hypothesis) until "proven" "guilty." The key difference from a trial is that we never conclude affirmatively that the alternative hypothesis is "true" in such terms.

Degrees of freedom is calculated by $(n_1+n_2)-2$, so it is $(40+65)-2=103$. Looking at Appendix B, we can see that there is no value for DF = 103. We will use the value for 120 for our estimate because it is the closest, even though it is just slightly less conservative. That value, our t_{crit} value, then, is 1.658. If t_{obs} exceeds 1.658 (in this case the difference should be positive), then we can reject the null hypothesis with some confidence and conclude that smokers use significantly more sick days than do nonsmokers.

We need the means (which we have), the sample sizes (which we also have), and...

We need variances for the two groups, but they're not given. Well, recall from before that the standard deviation is the square root of the variance. Therefore, the variance is the standard deviation squared. Simple. For the smokers, the variance is $0.98^2=0.9604$. For the nonsmokers, the variance is $0.77^2=0.5929$. All that gives us:

$$t = \frac{\bar{x}_1 - \bar{x}_2}{\sqrt{\frac{\sigma_1^2}{n_1} + \frac{\sigma_2^2}{n_2}}}$$

$$t = \frac{3.74 - 1.12}{\sqrt{\frac{0.9604}{40} + \frac{0.5929}{65}}}.$$

Just a refresher on order of operations, we need to go in a particular order: parenthesis, exponents, multiplication, division, addition and subtraction (PEMDAS). Here, we need to manage the business under the radical first, although in practical terms, we could do the subtraction in the numerator early on and just leave it alone until the proper time.

$$t = \frac{3.74 - 1.12}{\sqrt{0.024 + 0.009}}$$

$$t = \frac{3.74 - 1.12}{\sqrt{0.033}}$$

$$t = \frac{3.74 - 1.12}{0.1817}$$

$$t = \frac{2.62}{0.1817}$$

$$t = 14.42.$$

In terms of our *statistical decision*, our t_{obs}, then is 14.42. The t_{crit} was 1.658. 14.42 is considerably *larger* than 1.658, so we can defend rejecting the null hypothesis and conclude that smokers' use of sick time is significantly greater than that of nonsmokers.

Assumptions of Independent-Samples *t* Test:

Univariate normality

Homogeneity of variance

Truly continuous (metric) dependent variable.

The Snacky Statistician: Independent-Samples t Test Recipe Card

Basic Ingredients

One dependent (outcome) variable, continuous (metric) level of measure

One binary independent (predictor) variable

One hypothesis (one-tailed or two-tailed, depending on preference)

Directions

Whip up a mean and standard deviation for each group of the binary outcome variable (set aside variances for use in a moment).

Pool (average) the standard errors (using variances, not standard deviations) for the two variables. Use as denominator.

Find the difference between the two means. Make that the numerator.

Combine wet and dry ingredients according to the formula.

Compare result to t distribution at n-1 degrees of freedom.

Derive statistical decision.

Serve immediately.

Homogeneity of Variance

We can check into the degree of variance equality should we choose to do so. Some statistical software calculates a test, called Levene's Test for Equality of Variance. If Levene's test is significant, then the variances of the two groups is not sufficiently similar to justify the use of the normal pooled standard error. In such cases, we would use the simpler pooled standard error above.

If we are averaging variance for the two groups, it stands to reason that if one group contributes more variance to the pooled standard error, even if that average is a weighted average, comparisons between the groups can become problematic. For the same reasons, if two groups have wildly different sample sizes, then comparisons can become problematic. It deserves noting, though, that *t* test is rather robust against moderate deviations from variance equality (see, e.g., Cohen, 1988: 19).

Effect Size for Independent-Samples *t*

As we have observed, *t* tests are quite literally signal-to-noise ratios. It stands to reason, then, that (experimental conditions aside) the forces that influence our statistical decision are related to the amount of noise (by that I mean the amount of variability as measured by the standard error of the estimate) and also to the strength of the signal (what I called effect size earlier).

Regarding noise, both standard deviation and sample size contribute meaningfully to variance. Equality of variance is presumed, of course, but bigger samples make smaller standard errors if we keep standard deviation constant. Consider:

$$\sigma_{\bar{x}} = \frac{\sigma}{\sqrt{n}}$$

Let's say we have a standard deviation of 4.6. No reason. I just made it up. 4.6. If that standard deviation came from a sample $n = 15$, we would have:

$$\sigma_{\bar{x}} = \frac{4.6}{\sqrt{15}}$$

$$\sigma_{\bar{x}} = \frac{4.6}{3.873}$$

which is 1.1877. If, however, that same standard deviation came from a sample $n = 300$, things look rather different:

$$\sigma_{\bar{x}} = \frac{4.6}{\sqrt{300}}$$

which is

$$\sigma_{\bar{x}} = \frac{4.6}{17.321}$$

That would make the standard error 0.2656, which is only about 22% of the first standard error. As n increases, standard errors decrease; at some point, nearly *any* difference between the two sample means can be significant. Of course, that also means that if the two samples being compared had wildly different sample sizes… you guessed it: *t* test results can be weird.

Regarding the signal, we can express the size of the signal using several different measures discussed below.

Eta Squared and Point Biserial Correlation

One strategy for identifying the amount of signal present is to express the proportion of variance in the dependent variable that can be attributed to the independent variable. Eta squared (η^2) sets up yet another ratio:

$$\eta^2 = \frac{t^2}{t^2 + \text{df}}$$

Where t^2 is the t_{obs} from our t test squared, and degrees of freedom are, well, the appropriate degrees of freedom. The result is a number between 0 and 1, which indicates the percentage of overlap between dependent and independent variables: 0 indicates no relationship and 1 indicates a perfect relationship. If we calculate the square root of η^2, we have a **point biserial** correlation (r_{pb}), which is a measure of correlation between a binary variable and a metric variable (just like the variables we have in a t test). We will discuss point biserial correlation in Chapter 11.

$$r_{\text{pb}} = \sqrt{\frac{t^2}{t^2 + \text{df}}}$$

Cohen's d

Yet another measure of effect size is **Cohen's *d***:

$$d = \frac{\bar{x}_1 - \bar{x}_2}{s_{\text{p}}}$$

where s_p equals:

$$s_{\text{p}} = \sqrt{\frac{(n_1 - 1)(s_1^2) + (n_2 - 1)(s_2^2)}{n_1 + n_2 - 2}}$$

The s^2 terms above are the variances for each of the two groups, and n_1 and n_2 refer to the sample sizes. This, of course, looks *very much* like the denominator in the equal-variances-assumed pooled standard error (aside from the radical). Cohen's d offers an expression of the overlap between the curves of the two groups created by the independent variable. The larger d is, the stronger the signal is. The conventional interpretation of Cohen's d is that a score of 0.2 or below is "small," a score greater than 0.5 is "moderate," and a score greater than 0.8 is "large."

Paired-Samples t Test

In some cases, we are interested in the degree of change in something between two points in time, or between two matched groups. Although the reasoning is the same as it is for other t tests, both signal and noise must be derived rather differently because of the nature of the question.

$$t_\delta = \frac{\bar{D}}{\sigma_{\bar{x}_{\bar{\delta}}}}$$

I am using the Greek symbol δ or delta (which often indicates difference or change in something), and the capitol D in the numerator is also referencing a difference. The $\bar{D}$ is the mean change between time one and time two for each case; $\sigma_{\bar{x}_{\bar{\delta}}}$, then, is the standard error of that mean difference. That all means that we find the difference for each case, and then find the mean difference across all cases. The same is true for the standard error: we use the standard deviation of the difference, and the square root of the number of pairs.

$$\sigma_{\bar{x}_{\bar{\delta}}} = \frac{\sigma_{\delta}}{\sqrt{n_{\text{pairs}}}}.$$

Officer stress has interested (and concerned) scholars and leaders for some time (e.g., Storch & Panzarella, 1996). Recent events have made it manifestly clear that officer wellbeing (or lack thereof) can have dire consequences for citizens. It can also do considerable damage to officers. Cortisol is called "the stress hormone." Cortisol works to moderate the body's stress response, including adrenaline and insulin production (e.g., Cay et al., 2017). Prolonged high levels of cortisol can cause (or can be associated with) some rather serious consequences (e.g., Adam et al., 2017). Levels vary considerably over the course of the day, typically being quite a bit higher in the morning, and then sloping off toward bedtime. A normal morning cortisol level is between 10 and 20 micrograms per deciliter of blood (e.g., Chan & Debono, 2010).

A study was done with eight volunteer officers. For six months, volunteers' morning cortisol level was measured. After these six months, volunteers were given deep training in breathing meditation. Following a six-month period of practice, morning cortisol levels were taken for the same officers for six months.

In terms of hypotheses, we can think of things this way:

H_0: pre-practice cortisol levels *are not different from* post-practice cortisol levels
H_A: pre-practice cortisol levels *are different from* post-practice cortisol levels

Or

H_0: $\bar{D} = 0$
H_A: $\bar{D} \neq 0$.

After a thorough review of literature, Adam and colleagues (2017) determined that a flatter day-wise (or diurnal) cortisol curve was associated with poor mental and physical health in numerous studies, so it is not at all the case that we should assume a flatter diurnal curve is necessarily a thing to crave for our officers. It is not clear, though, whether the deleterious effects noted with flatter cortisol curves is cause or indicator. So too much cortisol over a long period is not the best, too little isn't fantastic, either, and a flatter curve is indicative of a variety of undesirable conditions. And again, we can't be certain whether cortisol changing is an indicator or cause of problems, or if it is acting as a mediator for "bad" stress (which we tend to think is corrosive on people, especially over time). What we can do in this circumstance, though, is examine changes in cortisol levels in relation to meditative practice; ideally, we would also include other measures of stress, including self-reported experiences of feeling stressed. Because we are not certain about the cortisol/health relationship (or at least I am not), a two-tailed hypothesis makes good sense.

If we use $p = 0.05$ for a two-tailed test at $n - 1$ degrees of freedom (eight officers, each with two scores... $n = 16$, so $n - 1 = 15$ degrees of freedom), our t_{crit} value is 2.131 (Table 9.1).

The next step is to find the difference between each measure. Table 9.2 shows the difference in column δ. The mean for that column is 3.26, and the standard deviation for these scores is 4.13.

Table 9.1 Pre- and Post-Practice Mean Morning Cortisol Levels for Officers

Officer	*Pre-Practice Mean Morning Cortisol (mcg/dL)*	*Post-Practice Mean Morning Cortisol (mcg/dL)*
A	13.5	13.1
B	27.9	18.8
C	19.7	14.5
D	18.5	19.3
E	30.3	20.8
F	34.3	33.5
G	21.7	19.9
H	18.3	18.2

Table 9.2 Pre- and Post-Practice Mean Morning Cortisol Levels for Officers with Pre–Post Difference

Officer	*Pre-Practice Mean Morning Cortisol (mcg/dL)*	*Post-Practice Mean Morning Cortisol (mcg/dL)*	δ
A	13.5	13.1	0.4
B	27.9	18.8	9.1
C	19.7	14.5	5.2
D	18.5	19.3	−0.8
E	30.3	20.8	9.5
F	34.3	33.5	0.8
G	21.7	19.9	1.8
H	18.3	18.2	.1

First, let's calculate our standard error:

$$\sigma_{\bar{x}_{\bar{\delta}}} = \frac{\sigma_\delta}{\sqrt{n_{\text{pairs}}}}$$

$$\sigma_{\bar{x}_{\bar{\delta}}} = \frac{4.13}{\sqrt{8}}$$

$$\sigma_{\bar{x}_{\bar{\delta}}} = \frac{4.13}{2.83}$$

$$\sigma_{\bar{x}_{\bar{\delta}}} = 1.46$$

So…

$$t_\delta = \frac{\bar{D}}{\sigma_{\bar{x}_{\bar{\delta}}}}$$

$$t_\delta = \frac{3.26}{1.46}$$

$$t_\delta = 2.233$$

Table 9.3 *t* Tests at a Glance

	Comparison	*Formula*	*Standard Error*	*DF*
Single sample	Sample mean and hypothetical population parameter	$t = \frac{\bar{x} - \mu}{\sigma_{\bar{x}}}$	$\sigma_{\bar{x}} = \frac{\sigma}{\sqrt{n}}$	$n - 1$
Independent sample	Two sample means (when the groups are thought to be different)	$t = \frac{\bar{x}_1 - \bar{x}_2}{\sigma_{\bar{x}p}}$	$\sigma_{\bar{x}p} = \sqrt{\frac{\sigma_1^2}{n_1}} + \sqrt{\frac{\sigma_2^2}{n_2}}$	$(n_1 + n_2) - 2$
Paired sample	Two sample means (for connected or dependent comparisons – e.g., cohort or panel studies)	$t_\delta = \frac{\bar{D}}{\sigma_{\bar{x}\delta}}$	$\sigma_{\bar{x}\delta} = \frac{\sigma_\delta}{\sqrt{n_{\text{pairs}}}}$	$n - 1$

In terms of our *statistical decision*, then, if t_{crit} is 2.131, and our t_{obs} is 2.233, we can reject the null hypothesis that the difference between pre- and post-practice is zero. Not by much, and we are not beholden to do so, but we can reject H_0. It stands to reason, though, that we should have some reason for our decision.

Summary

Student's *t* test is a capable mechanism for performing bivariate (or in the case of single-sample *t* tests, univariate) hypothesis testing. Single-sample *t* tests are used when we seek to compare a sample mean to a hypothesized value. Independent-samples *t* is the choice for when we look to compare two independent, parallel samples to each other. The paired-samples *t* test is the best option for when we want to determine if there has been a change over time, or if there is a difference between matched pairs of cases. Assumptions for the tests, including a true metric nature for the dependent variable, and homogeneity of variance, are important considerations when using *t* tests.

Main Take-Aways

- Student's *t* test allows us to examine a metric dependent variable in terms of a binary (two-level) independent variable.
- Student's *t* test is our first contact with a technique that utilizes the reasoning that has come before, including standard error, null and alternative hypotheses and a statistical decision.
- There are three kinds of *t* tests, and each has a particular setting in which it is the appropriate choice. Although differences in the numerator are obvious, one of the chief differences among the three types of *t* tests is how standard error is managed. Standard error for independent-samples *t*, for example, is a weighted average of the standard errors for the two variables.

Key Terms (in order of appearance)

single-sample *t* test
effect size, independent-samples *t* test
Cohen's *d*
independent-samples *t* test
η^2 (Eta squared)
paired-samples *t* test
homogeneity of variance
point biserial

Chapter 10

Bivariate Hypothesis Testing

ANOVA

Introduction

In Chapter 9, we examined a strategy – three related strategies, in fact – useful to determine a kind of meaningfulness; t tests allow us to understand the importance, in terms of relative likelihood, of the distance between two means of different kinds. They are robust and useful techniques that are often overlooked in favor of fancier multivariate techniques. They are limited in scope, however; in a circumstance where an independent variable has more than two values, the t test becomes problematic, and rather quickly, in fact. Take an example where a metric dependent variable has only three discrete values. In order to compare means for each of three groups, three comparisons would be needed. If that same variable had four values (called levels), *six* t tests would be needed. If it had five values, we would need *ten* separate t test comparisons.

Alpha Inflation

OK, that might just seem like a problem of more calculation, fully solvable with computers, but something more insidious is at work here. Recall how statistical significance works: by performing analyses on a sample derived in random (read EPSEM) fashion. Living in accord with the Central Limit Theorem, we would return to the population for another random sample for *each test* we ran. Better still, we would return to the population repeatedly (with replacement) for one test in order to derive a truly precise estimate of μ. Instead of drawing multiple *samples* for each *test*, what is proposed above is running multiple *tests* on the same *sample*. Recall that statistical significance is determined when an outcome is found *to be possible by chance, but only rarely*. If we make repeated inferences from a single sample, we strain the reasoning of the Central Limit Theorem. A sampling distribution, recall, is a collection of unbiased means drawn from a population. Any one sample, at least in theory, can support only a limited amount of conclusion-drawing on the basis of the Central Limit Theorem. If any one t test produced a significant finding, $p = 0.05$, that means that the difference found could happen by chance only one time in 20. If we did three separate t tests, as we would have to do for an independent variable with three discrete values, if each used the $p = 0.05$ standard, the probability of a significant finding is not one time in 20, but is *three* in 20, or 0.15 (ahh, there's that addition rule again). If the independent variable had four levels, the six comparisons required, each at $p = 0.05$, would make alpha 0.3, and if the variable had five levels, the probability of a significant finding *by chance alone* becomes 0.5, or *one chance in two*. Uh oh.

DOI: 10.4324/9781032638591-10

The phenomenon described above is called **alpha inflation**; when we run multiple tests from the same sample, we risk rejecting the null hypothesis inappropriately (Type I error). To correct for alpha inflation, we could lower the alpha level for each individual test, but that can have serious consequences by increasing the likelihood of Type II error (if the *p* value is excessively small, like 0.01 or smaller, it will be only truly extremely unlikely outcomes that will be determined to be significant). As a solution, decreasing per-test alpha across all planned tests is a strategy, actually it is a strategy we call Bonferroni correction, but that's something we will discuss in a moment.

No, a markedly better solution for a single hypothesis with an independent variable with multiple values (or levels) is to use an omnibus comparison that seeks to fold in comparisons for means of each group into a single test. We have just such a technique: analysis of variance or ANOVA.

But wait… you might clench your fist and rage at the choice of name for the test. "It is an analysis of *means*, you scalawag!!" I can hear you shout to the sky through gnashed teeth. And you'd be right, except like with *t* test, ANOVA compares means *in the context of group variances*; it, like *t* test, is a signal-to-noise ratio.

It is called an analysis of *variance*, though, because although it compares means, it uses variances to make the comparisons. The ratio that it creates is, in its most fundamental sense, a ratio of variances, specifically, the ratio can be understood as:

$$F = \frac{\text{between group variance}}{\text{within group variance}}$$

ANOVA uses a single dependent metric variable and a single categorical independent variable (often called a factor) and seeks to determine whether the levels of the factor differ significantly with regard to the dependent variable; it asks about the degree to which the metric dependent variable is dependent upon group (or factor) membership for each level of the independent variable (or factor).

As with *t* test, there are multiple versions of ANOVA, depending on the particular nature of the comparisons to be made. We can think of ANOVA as a class of techniques, each with its own applicability. If a single dependent variable is examined across levels of a single factor (that is to say, if means for the dependent variable are compared across different groups in the independent variable), then we do a **one-way between-S ANOVA**. We might, for example, look at the professionalism index GPA from before, and see if there are differences in the GPA mean for racethnic groups in the sample. In such a circumstance, we would be seeking to determine if the professionalism index GPA varied meaningfully by racethnic category. We say that it is "between-S" to indicate that it is making comparisons between (really, or among) samples. If, as with a paired-sample *t* test, we wanted to compare pre- and post-intervention measures, or if we had matched samples, we could use the imaginatively named **repeated-measures ANOVA**. If our model used more than 1 factor, that is, that it involved more than 1 categorical independent variable, we would do **factorial ANOVA**. This chapter will focus only on one-way between-S ANOVA, leaving other members of the analysis of variance family for other treatments.

Like with *t* test, ANOVA operates from some rather particular assumptions. In fact, the assumptions about the data and the model are for all practical purposes the same as the assumptions for *t* test: univariate and bivariate normality, homogeneity of variance across groups and a truly metric dependent variable.

The null hypothesis for an ANOVA is that the means of each of the *k* populations (operationalized as the *k* levels, or groups, in the independent variable) are the same:

$$H_0: \mu_1 = \mu_2 = \mu_3 \cdots = \mu_n$$

Instead of using a *t* or *z* distribution, ANOVA uses an *F* distribution. *F* distributions, like *t*, vary in shape according to degrees of freedom (but degrees of freedom for both numerator and denominator separately); instead of a symmetrical bell-shape like *z* or *t*, *F* (really, the many *F*s) has (or have) a clear positive skew. It is a little different for *F* than it is for *t*, but that's still to come. Appendix C presents the values of the *F* distribution.

The Basic Logic of ANOVA

Figure 10.1 is identical to Figure 9.1 because it uses the terms appropriate to both *t* test and ANOVA (OK, I thought ahead just a little). The between-group difference, or signal, is that difference between-group means, and the within-group variance is the rattle-and-hum within each group. The ratio ANOVA crafts is, like in *t* test, a ratio of between-group variance (signal) to within-group variance (or noise). The only real difference between *t* test and simple one-way ANOVA is the number of groups. If we added a third, fourth or more groups (more bell-shaped distributions) in Figure 10.1, we would be doing ANOVA, not *t* test.

Although the reasoning is the same for ANOVA as it is for *t* test, the way which signal and noise are computed differ considerably. With *t* test, of course, the signal in the numerator is just the difference between two means, or between a mean and a hypothetical population parameter; only subtraction is needed to find the signal (the amount of difference). For ANOVA, precisely because it isn't just a comparison between two means, we need to measure signal in a more comprehensive way.

The process for measuring signal, though, is a familiar one. Recall the calculation of standard deviation? We found the mean, and then compared each score to the mean, and then squared that difference. Well, we could call that process the sum of squared errors, or

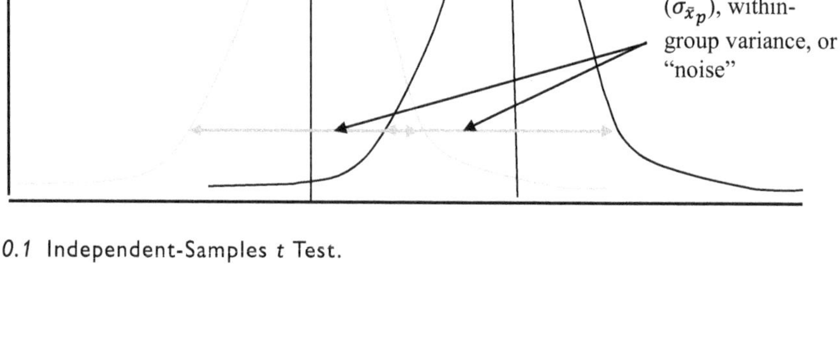

Figure 10.1 Independent-Samples *t* Test.

the **sum of squares** (often denoted as SS). This process, summing squared errors, is the very process we can use to produce a measure of signal.

Actually, we do well to think about ANOVA as a ratio of two similar proportions, *almost* a ratio of two signal-to-noise ratios. Each proportion, the numerator and the denominator, is composed of a measure of difference with a measure of context. Specifically, in both the numerator and the denominator, we combine a sum of squared errors (or sum of squares) and a measure of degrees of freedom. The basic shape of ANOVA, then, looks like this:

$$F = \frac{\text{Sum of squares between/degrees of freedom between}}{\text{sum of squares within/degrees of freedom within}}$$

When we combine the sum of squares and degrees of freedom, though, we create a term we call the **mean square.** So, ANOVA is essentially:

$$F = \frac{\text{mean square between}}{\text{mean square within}}.$$

The numerator, our expression of "signal," is an expression of the variance of differences among group means from the grand mean of the total sample, contextualized by degrees of freedom for the term. The mean square within – our "noise" – is an expression of the variance within each of the groups being considered, contextualized by the degrees of freedom within which that variance can operate across all groups.

One-Way, between-S ANOVA

As I said already, one-way ANOVA is used when we have a single metric dependent variable and a categorical independent variable (a single factor) with multiple groups (levels). The calculation of ANOVA is quite simple mathematically, but is somewhat tedious, both in terms of the many steps, and in terms of the many calculations for a number of the steps. Recall from above that we have four basic measures to derive:

1 Sum of squares between (also called the between-group sum of squares)
2 Sum of squares within (or, imaginatively, the within-group sum of squares)
3 Degrees of freedom between
4 Degrees of freedom within

Let's discuss each in turn.

Between-Group Sum of Squares

For our measure of signal, what we need is a comparison of each group mean (symbolized as $\bar{x}_k$) to the grand mean (that is, the mean for the entire sample, symbolized as $\bar{\bar{x}}_Y$). Doing so gives us a comparison for each group to what might otherwise be expected as a value for the group mean. In order to determine the between-group sum of squares, we need just a few measures: the mean for each group and the grand mean. First, we find each group mean (we add up all group scores and then divide by the number of scores in the group), then we find the grand mean across all scores regardless of group. Then we would find the difference from the grand mean for each group mean (again, using subtraction), then square that difference and then just add them all up. Just that simple. Conceptually, this expression of signal is not all that different from how we create a measure of signal in independent-samples and paired-samples *t* tests. The formula looks like this:

$$SS_{\text{Between}} = \Sigma\left(\bar{x}_k - \bar{\bar{x}}_Y\right)^2$$

Within-Group Sum of Squares

Almost as simple, but rather more involved, the within-group sum of squares (our expression of noise) is derived by comparing the score for each group member to the appropriate group mean ($\bar{x}_k$), and then adding them all up for each group. What we are seeking is the amount of rattle-and-hum within each group, so that we can pool it. That pooled (again, or averaged, when we divide by degrees of freedom) rattle-and-hum lets us know about how much of a difference we might expect to exist *in situ* due to normal variation of the thing we've measured with our dependent variable. And that'd be:

$$SS_{\text{Within}} = \Sigma\left(x_i - \bar{x}_k\right)^2$$

Degrees of Freedom Between

In order to contextualize how much variation there *is* between-group means, we need to know how much variation there *could be* between those same means. The degrees of freedom between groups is $k - 1$, or the number of groups minus one.

Degrees of Freedom Within

In order to contextualize how much variation there *is* within-group means, we need to know how much variation there *could be* between those same means. The degrees of freedom within groups is $n - k$, or the number of total *cases* in the sample minus the number of *groups*.

Perhaps it is time for an example. A study was done that compared two stress-related interventions for officers with a comparison group that involved a stress management class. Again, morning cortisol levels were measured; post-intervention levels are presented in Table 10.1.

So, our null hypothesis looks like this:

$$H_0: \mu_a = \mu_b = \mu_c$$

Or

$$H_0: \mu_x - \mu_y = 0$$

Meaning that the difference between any of the three (in this case) means is 0.

Our alternative hypothesis looks like this:

$$H_A: \mu_a \neq \mu_b \neq \mu_c$$

Table 10.1 Officer Cortisol and Intervention

Officer	*Group*	*AM Cortisol (mcg/dL)*
A	Meditation	12.1
B	Meditation	13.2
C	Meditation	10.9
E	Workout	14.1
F	Workout	15.5
G	Workout	14.7
H	Stress Management Class	16.5
J	Stress Management Class	15.6
K	Stress Management Class	17.0

Or that the differences between the groups are nontrivial, that they are not the same. Said another way, the three groups do not appear to belong to the same parent population. Again, should we find in favor of the alternative hypothesis, we would not conclude that it was "true," but instead merely that the null hypothesis could be rejected as being valid for this sample.

So, the first thing to do is to acquire the needed means. We need a grand mean, or a mean cortisol level for all nine participants. The grand mean ($\bar{\bar{x}}_Y$) is 14.4.

Then we need means for each group. The mean for the meditation group is 12.067; the mean for the workout group is 14.767 and the mean for the stress management class (SMC) is 16.367. Now, we can see that there are differences, but the question we are asking is whether these differences are attributable to chance, or if something else is at work, evidenced by the differences in means. Again, we can't know that there's a substantive difference from statistical analyses. We can *infer* one based on how likely the difference we see is given the context of natural variation, or rattle-and-hum, in the variable itself.

Next, we need to make our comparisons. First, let's do the between-group sum of squares.

If we add up all the numbers in the far-right column of Table 10.2 we get 28.34, which is our between-groups sum of squares. Next, we need to calculate our within-group sum of squares.

To be clear, the last column in Table 10.3, the per-group sum of squares isn't necessary, but it is added to emphasize the grouped shape to the within-group comparisons; recall that the assumption of variance equality is present in ANOVA as it is in *t* test, and eyeballing the group means is an early look into whether that assumption makes sense. Because we will divide that sum later by degrees of freedom, it is pooled, or averaged within-group deviation. We could have as easily simply summed the values in the Squared Difference column, but we would have missed the chance to look at groupwise influence on the pooled within-group sum of squares. Anyway, we have our within-group sum of squares: 4.638. What remains is to derive the measures for degrees of freedom and then calculate our *F* ratio.

First, let's do our degrees of freedom between, calculated via $k-1$. In our example, there are three groups (meditation, workout and SMC) so $k=3$. That means that $k-1=2$. Easy. Now, let's do the degrees of freedom within. We have nine cases ($n=9$) and three groups. The formula for degrees of freedom within is $n-k$, so, $9-3=6$.

To review, then, our between-group sum of squares is 28.34 and our between-group degrees of freedom is 2. Our within-group sum of squares is 4.638 and our within-group

Table 10.2 Between-Group Sum of Squares

Officer	*Group*	*AM Cortisol (mcg/dL)*	*Group Means* ($\bar{x}_k$)	*Grand Mean* ($\bar{\bar{x}}_Y$)	*Difference*	*Squared Difference*
A	Meditation	12.1	12.067	14.4	−2.333	5.444
B	Meditation	13.2		14.4	−2.333	5.444
C	Meditation	10.9		14.4	−2.333	5.444
E	Workout	14.1	14.767	14.4	0.367	0.134
F	Workout	15.5		14.4	0.367	0.134
G	Workout	14.7		14.4	0.367	0.134
H	SMC	16.5	16.367	14.4	1.967	3.868
J	SMC	15.6		14.4	1.967	3.868
K	SMC	17.0		14.4	1.967	3.868
Σ:						28.34

Table 10.3 Within-Group Sum of Squares

Officer	*Group*	*AM Cortisol (mcg/dL)*	*Group Mean* ($\bar{x}_k$)	*Difference from Group Mean*	*Squared Difference*	*Group SS*
A	Meditation	12.1	12.067	0.033	0.001	2.646
B	Meditation	13.2		1.133	1.284	
C	Meditation	10.9		−1.167	1.361	
E	Workout	14.1	14.767	−0.667	0.444	0.985
F	Workout	15.5		0.733	0.537	
G	Workout	14.7		−0.067	0.004	
H	SMC	16.5	16.367	0.133	0.018	1.007
J	SMC	15.6		−0.767	0.588	
K	SMC	17.0		0.633	0.401	
Σ:						4.638

degrees of freedom is 6. We have what we need to proceed. Our formula again for the F ratio in ANOVA is:

$$F = \frac{\text{sum of squares between/degrees of freedom between}}{\text{sum of squares within/degrees of freedom within}}$$

So...

$$F = \frac{28.34/2}{4.638/6}$$

Which is

$$F = \frac{14.17}{0.773}$$

$$F = 18.331$$

Our mean square between is 14.17 and our mean square within is 0.773. That means our F ratio is 18.331. That is our F_{obs}. Of course, we now need our F_{crit} in order to make our statistical decision regarding the hypothesis. Referring to Appendix C, along the top, we can find the degrees of freedom ("df") for the numerator (2), and along the side, we can locate the degrees of freedom for the denominator (6). The corresponding critical value at $p=0.05$ is 5.14. That would mean that our F_{obs}, 18.331, being greater than the F_{crit}, falls in the rejection region, and so we have reason to reject the null hypothesis and conclude that the difference in group means is nontrivial statistically.

Now there's another point to be made here. You might've caught it already, especially if you've looked online at videos on how to cook an ANOVA. We have talked about within-group and between-group sums of squares, but there's a third kind. If we were to take each score and subtract it from the grand mean, and then square the difference and sum it all up, we would have the total sum of squares. That's the total amount of variance kicking around in the measurement.

Partitioning of the Sum of Squares

Now it just so happens that if we were to add up the between-group sum of squares and the within-group sum of squares, just as we would have done for calculating a standard deviation, we would get $28.34+4.638=32.98$. That's the same as the total sum of squares as seen in Table 10.4.

$$SS_{total} = SS_{Between} + SS_{Within}$$

Just for the sake of stuff coming later, another way to say the same thing is:

$$SS_{total} = SS_{Regression} + SS_{Residual}$$

The between-group (regression) sum of squares (also known as "signal") and the within-group (residual) sum of squares (the "noise") will always sum to the total sum of

Table 10.4 Total Sum of Squares

Officer	*Group*	*AM Cortisol (mcg/dL)*	*Grand Mean* ($\bar{\bar{x}}_Y$)	*Difference*	*Squared Difference*
A	Meditation	12.1	14.4	−2.3	5.29
B	Meditation	13.2	14.4	−1.2	1.44
C	Meditation	10.9	14.4	−3.5	12.25
E	Workout	14.1	14.4	−0.3	0.09
F	Workout	15.5	14.4	1.1	1.21
G	Workout	14.7	14.4	0.3	0.09
H	Stress Management Class	16.5	14.4	2.1	4.41
J	Stress Management Class	15.6	14.4	1.2	1.44
K	Stress Management Class	17.0	14.4	2.6	6.76
Σ:					32.98

squares. In essence, if we think of the total sum of squared differences (or error) in Table 10.4 as the total amount of variability, we are dividing (or **partitioning**) that total amount of error into two piles: the first pile is that amount of variation that is attributable to the between-group differences, or the signal, and the second pile is the amount of difference attributable to the within-group differences, or the noise. The partitioning allows us to create a ratio of the variability attributable to group membership to the variability attributable to error (or what I've called rattle-and-hum throughout this text).

ANOVA Calculation at a Glance

The total sum of squares is calculated by:

$$SS_{\text{Total}} = \Sigma\left(x_i - \bar{\bar{x}}_Y\right)^2$$

Or the squared sum of each score (x_i) minus the grand mean $\left(\bar{\bar{x}}_Y\right)$.

Again, the between-group (or regression) sum of squares is calculated by:

$$SS_{\text{Between}} = \Sigma\left(\bar{x}_k - \bar{\bar{x}}_Y\right)^2$$

For each score, we find the difference between the group mean $\left(\bar{x}_k\right)$ and the grand mean $(\bar{\bar{x}}_Y)$, and then square each difference and add them up.

The within-group (or residual) sum of squares is calculated:

$$SS_{\text{Within}} = \Sigma\left(x_i - \bar{x}_k\right)^2$$

For each score, we find the difference between that score and the mean for its group $\left(\bar{x}_k\right)$, square the difference and then add them up.

Between-group degrees of freedom is $k-1$, and within-group degrees of freedom is $n-k$. Then we make the final ratio:

$$F = \frac{\text{sum of squares between/degrees of freedom between}}{\text{sum of squares within/degrees of freedom within}}$$

Post-Hoc Analyses

Having made our central statistical decision, it remains for us to consider a number of conditions relative to the analyses. Our statistical decision is not the only thing we can glean from our analyses of the relationship under consideration, and in fact, in order to understand the particulars of that relationship, there are several analyses we can perform. Often, these post-hoc (after-the-fact) or **"protected" analyses** give us richer and more nuanced contextual insights into the relationship under investigation. They're called "protected" because they dwell on the far side of the statistical decision, isolated from the Central Limit Theorem's doctrinaire purity. They are handy apostates, wantonly shredding some of the inferential assumptions we have spent so long detailing up to this point, sometimes even laughing while they do so.

OK, that might be a bit over the top.

Anyway, one post-hoc analysis we can perform is related to the effect size. If you recall from Chapter 9, we talked about η^2 as an expression of effect size for *t* test. We can use the same test, calculated even more simply, for determining effect size in ANOVA. Very simply, η^2 is calculated:

$$\eta^2 = \frac{SS_{\text{between}}}{SS_{\text{total}}}$$

Of course, this only stands to reason; it tells us how big the effect (or signal) is relative to the total amount of variability. As it is with *t* test, η^2 is a proportion, expressed as a decimal between 0 and 1; if it is 1, then all of the variability is signal. If the number is at or near 0, then none of the variability is properly understood to be signal.

Another technique for post-hoc analysis is, in essence, to do the very *t* test-like pairwise comparisons we avoided in the first place by using ANOVA. Tukey's honestly significant difference uses ... well, it uses a *t* test, actually:

$$q = \frac{\bar{x}_{\text{a}} - \bar{x}_{\text{b}}}{\sqrt{MS_{\text{within}}/n}}$$

Yep, it is a straight-up *t* test of a kind, with the MS_{within} serving as a standard error. The q is used by Tukey to indicate that the *t* distribution is not to be used for the determination of significance. Instead, the Studentized Range Statistic table would be used (not included in this text).

What is happening here is a closer examination of group means relative to each other; sometimes a researcher will examine the group means in advance and set up particular group contrasts in advance in order to prevent alpha inflation, but again, if the comparisons are made post-hoc, the possibility of alpha inflation is, at least theoretically, set aside.

Bonferroni Correction

A rather more conservative alternative is to use **Bonferroni correction.** Bonferroni correction is a strategy for reducing the risk of alpha inflation when we perform multiple inferential tests:

$$PC_{\alpha} = {}^{EW_{\alpha}}\!/_{k}$$

Or the per-test alpha level is the total experiment-wide alpha divided by the number of tests to be performed (k). It really is just that simple. Of course, it isn't fantastically good news; if we perform four tests as part of the same experiment, and we want $p=0.05$ to be our experiment-wide alpha, that means that our p value for any given test, the critical threshold for the determination of significance, is now 0.0125 per test. That's not very high, and it tends to inflate the probability of Type II error. This is why ANOVA is such a useful technique, even when we follow Tukey's advice and make the pairwise comparisons that would necessitate Bonferroni correction in the first place.

Summary

Although the three *t* tests we just discussed are capable devices for hypothesis testing under proper circumstances, if we have an independent variable with more than two levels, we need to rethink some things. We can run multiple *t* tests, but doing so rather quickly imperils us with regard to the likelihood of Type I error. Instead of basing hypothesis tests on each of several *t* tests, we can perform an analysis of variance. Notably, there are several members of the analysis of variance family, but we will discuss only one of these here: the one-way, between-S analysis of variance.

Main Take-Aways

- Multiple repeated tests from the same sample, like doing several *t* tests, comes at a cost – Type I error becomes increasingly likely with the more tests one does on the same sample.
- Addressing that cost can be done in a few ways. One way is to perform Bonferroni correction, but the Bonferroni adjustment represents a material increase in the likelihood of Type II error in order to manage the risk of Type I error.
- Bonferroni correction is a fine alternative in some settings, but another option may be more appropriate: an omnibus test like one-way ANOVA.
- ANOVA represents another, perhaps clearer example of the signal-to-noise principle that has been a key point of this book. ANOVA works like *t* test in that it takes a score, itself a ratio of a kind, and compares it with a probability density function. In the case of ANOVA, we use the F distribution for our purposes.
- ANOVA can be understood as a ratio of ratios: the result of the sum of squares divided by the between-group degrees of freedom slots in as the numerator, and the result of the sum of squares divided by the within-group degrees of freedom as the denominator.
- In order to interpret ANOVA, one trick is to do the very multiple *t* tests we sought to avoid for fear of upsetting the Central Limit Theorem, but we do that after the statistical decision has been made, so the risk of alpha inflation is limited, at least theoretically.

Key Terms (in order of appearance)

alpha inflation	one-way, between-S ANOVA	repeated-measures ANOVA
factorial ANOVA	sum of squares	mean square
between-group sum of squares	within-group sum of squares	degrees of freedom within
degrees of freedom between	partitioning of sum of squares	post-hoc analysis
protected analysis	Bonferroni correction	

Chapter 11

Bivariate Hypothesis Testing

Measures of Correlation

Introduction

Up to this point, we have discussed comparisons with a single metric variable and a categorical variable. In the main, the comparisons have examined the values of that one metric variable in terms of its variability across levels of the categorical independent variable. These are wildly robust and useful analytic strategies, but it is fair to some degree to think of them as operating on unequal terms; two variables are involved, so it is proper to think of them as bivariate techniques (except for single-samples *t* test, as I've mentioned), but the two variables do not work as equals in the analyses. The nuanced variation that drives the analyses is almost exclusively a function of the metric dependent variable, whereas the class membership of the independent variable tends to be the thing that really interests us. Most of the calculation is done in terms of the dependent variable, so as to understand something about the meaning of the different levels of the independent variable. That's all about to change, though; the techniques we will discuss in this chapter contemplate two variables in terms of how they interrelate and they will do so to some degree on relatively more equal terms with regard to the calculations we use.

A couple of the measures we have calculated already seek to treat two variables on more equal footing; point biserial, discussed very briefly in the context of *t* tests, is an example and will be treated in greater detail here. The tests we will discuss in this chapter involve the consideration of the nature of the relationship between two variables. More particularly, we will be examining the degree to which two variables, like dancers on a dance floor, appear to move together.

Recall how when we talked about causation, we indicated there were three things we needed? Those things were:

1 Correlation
2 Temporal Order
3 Nonspuriousness

Well, the first one of these, correlation, is just this co-movement, or co-relation. Oftentimes, researchers will designate one variable *X* and another as *Y*, implying an independent–dependent variable relationship, but outside of a laboratory setting, inferring causation from a correlation analysis is improper. If laboratory conditions can *definitively* isolate spurious influences and design conditions can ascertain temporal order, then correlation would be the last thing needed to create a convincing argument for causation (e.g., see Warner, 2008).

DOI: 10.4324/9781032638591-11

As it turns out in all but the most extraordinary circumstances, though, criminological conditions do not permit for methodological isolation to the degree necessary to derive causative conclusions from a simple correlation.

There's a story about a carpenter who owned only a hammer. Every project for him, because of his limited kit, was understood in terms of hammering. This apocryphal cautionary tale-in-brief is not a critique of hammers, but instead it is a warning to us to derive solutions for problems from the problems themselves, and not in terms of the solution-like things we have in ready reach. There are certain statistical tests that are, as the kids today say, "iconic." Linear regression is one of these, and Pearson's r is also one (as we will see, these two techniques are very closely related). These tests, perhaps because of their intuitive appeal, have had an outsized impact on researchers' thinking and even how other tests are crafted and interpreted. The problem with all of this, though, is that we tend to operate in our world such that things in that world fit into the frameworks that tests like these tend to provide. Doing so can diminish the fertility of the research we do that seeks to understand that world in nuanced and elegant ways.

The Logic of Correlation Analyses

One reason correlation analyses are so influential is that they allow for the consideration of a question that is often an intuitive one: to what degree do two measured things seem to move together. Do they move together like soldiers in tight lockstep, perhaps like wingborne swallows or agile baitfish moving in sudden, graceful unison, or do they move together almost not at all? Maybe they move together in clumsy, partial fashion, or even with one lagged behind the other? Another reason for the influence of correlation analyses, particularly Karl Pearson's r is that the resulting number, the number that we use for interpretation (which we might think of in terms of the read-out from the analysis) is also exceedingly intuitive. Pearson r scores fall between +1.0 for perfect direct relationships (where X and Y both increase in perfect unison) and –1.0 for perfect inverse relationships (where one variable decreases in perfect unison with unit increases in the other variable). When two things move together in extremely tight ways, Pearson's r is at or near +1.0 or –1.0. If there is discernible co-movement, but that movement is only partially in unison, the Pearson r score will be moderate, somewhere in the 0.5 (or –0.5) range. If the two variables do not move together at all, like strangers in different parts of the world, one fast asleep and one dancing in a wedding, then Pearson r is at or near 0.

Just intuitively, you might have assumed already that the two variables involved in most correlation techniques are metric, and in the case of Pearson r, you'd be quite right. Recall that point biserial is a measure of correlation between a metric variable and a binary categorical variable. There are tests like point biserial that don't involve comparisons between variables with similar levels of measurement, but even they are understood in terms that are similar to the gold standard Pearson r. Yet other measures of association accommodate ably variables with ordinal or even nominal levels of measure.

It is useful to point out at this point that with other tests so far, typically, we run one or a handful of tests (we do an ANOVA or a t test) at a time. With Pearson correlation at least, we will ask statistical software to produce something called a **correlation matrix.** The matrix presents correlations with each of the two bivariate pairings presented in the table. Now, I imagine some of you might be a tad wary of the idea of running multiple tests at once because of the risk of alpha inflation, and again, you would be most right to be so;

Table 11.1 Correlation Matrix with Ten Random Variables

	Var1	*Var2*	*Var3*	*Var4*	*Var5*	*Var6*	*Var7*	*var8*	*var9*	*Var10*
Var1	1.00									
Var2	0.13	1.00								
Var3	−0.27	0.22	1.00							
Var4	0.39	0.17	−0.66	1.00						
Var5	−0.14	0.65	0.18	0.14	1.00					
Var6	−0.08	−0.11	0.27	−0.08	−0.03	1.00				
Var7	−0.03	0.41	0.01	0.00	0.79	−0.29	1.00			
Var8	0.02	0.26	0.64	−0.35	0.02	−0.04	0.18	1.00		
Var9	−0.21	−0.40	0.12	−0.25	−0.05	−0.25	−0.16	−0.27	1.00	
Var10	−0.03	0.32	−0.27	−0.21	0.00	−0.25	0.07	−0.19	0.05	1.00

just on the basis of statistical reasoning, if we compare ten variables with randomly generated values for each of ten cases (Table 11.1), we would be making *45 individual* bivariate comparisons. Just on the basis of random chance, if $p = 0.05$, then we should expect 2.5 comparisons to be significant just due to chance alone. Let's take a look and see.

We had just a little more than the 2.5 comparisons turn out to be significant (the p values are not shown here in the table). We expected 2.5 (really, 3, practically speaking) but got 4. That difference from expectation is notable, but these results certainly make the point that any correlation matrix with numerous variables will produce significant relationships *with or without the presence of any substantively meaningful relationships.*

Now, we should speak about the determination of statistical significance for Pearson r. I did not include tests of significance for Table 11.1, but *conservatively*, the scores circled might credibly turn out to be statistically significant. We might've expected two to be significant (in this case, we are speaking in terms of magnitude of r, not of the actual p value, although they are related). More, there are several others that are likely candidates for statistical significance (values greater than + or −0.2 are defensible candidates), depending on sample size and nature. The next task would be to interpret the values we have in the context of expectations from theory. If they make sense, they might well be meaningful. If they don't, they may be spurious. Of course, if they don't make sense, they may still be accurate reflections of a reality in the parent population. More, that they make sense is in no way a guarantee that they're substantively meaningful. The correlation matrix shown in Table 11.1 would be seen as a decent outcome for a human sciences project, especially if variables 2, 5, 3, and 8 were thought at the outset to be related somehow. Bonferroni correction (discussed in Chapter 10) can be a strategy for reducing the danger of alpha inflation here, too, but it comes with the same costs as it does elsewhere.

Next, let's make note of the anatomical points for our correlation matrix. First, you can see that the top right triangle is missing. That's because it would be redundant to the values in the bottom left triangle (the comparison of Var7 and Var9 is the same value as the comparison of Var9 and Var7, so we only really need one of them). Further, the 1.00 indicates the perfect correlation between each variable and itself. We can ignore that, as it is to be expected.

The main take-away from Table 11.1, then, is that we should be wise with regard to the number and nature of comparisons we make, and that *a priori* theory is a useful (though not fool-proof) guide for the interpretation of results.

Let's talk about each of several different kinds of correlational measures of association.

Pearson's *r*

Just like other tests we have discussed thus far, Pearson *r* operates with some rather specific assumptions:

1 Variables should be independent measures of phenomena;
2 *X* and *Y*, the two variables, should be related in some kind of linear fashion;
3 Both variables should be interval or ratio (that is, metric) levels of measure;
4 Scores are paired, meaning that for each case (say, for each person in our data), there are two valid scores, one each for *X* and *Y*;
5 *X* and *Y* should both be univariate and bivariate normal in their distributions, and no significant outliers are present.

Let's discuss each in turn.

Independence

We can think about independence in terms of the purity of the comparisons we seek to make. Influences beyond the variables we seek to combine in analyses are the concern of independence. Another way of saying this is that errors (again, not mistakes but the differences between predicted scores and actual scores) are not correlated. *Yet another* way of saying this is that the scores of *X* and *Y* for a given case are not related in some way to the scores of *X* and *Y* for another case. Recency or proximity of observations is a common source of influence that violates independence. If I'm reading your survey over your shoulder as you answer it and I answer it sometimes as you do (because I think you're cool, of course), I create an independence problem. Similarly, if I take a survey, and then take it again a week or so later, can remember my answers and respond from memory, I create an independence problem. And of course, I am confounding the precision of the measure, thus impacting internal validity.

Let's say we wanted to compare criminal history (number of convictions), which we will call *X*, with current sentence in days, which we will call *Y*. It stands to reason that *X* and *Y* are related to some degree. If, though, previous convictions are recent events, say many came from participation in protests, then the measure for criminal history is not independent.

Linearity

Well, to be fair, Pearson's *r* is actually a test to measure the degree of linearity of the relationship between two variables. If the relationship isn't linear, then other tests are more appropriate (on which I will wax poetic shortly). So, from a pure perspective, Pearson doesn't just *assume* linearity; it *tests* for linearity. We do well for ourselves, though, if we eyeball for linearity to begin with, and then use Pearson *r* if linearity seems evident.

Figure 11.1 illustrates linearity in the form of a scatterplot of our variables *X* and *Y*, complete with a trend line that portrays the linearity of the relationship. To be sure, the dots don't all fall on the line, but there does not appear to be a great deal of variation of errors in *Y* across values of *X*. In contrast, consider Figure 11.2: the same line seems to fit the general slope of the *X*/*Y* relationship, but clearly, the line seems to miss scores systematically at certain places over the arc of *X* values. Specifically, if our trend line was an estimate of any given value of *Y* given *X*, it guesses systematically higher on *X* for low and high values of *Y*, and systematically lower for mid-ranged values of *Y*. The Pearson *r* for the two variables in Figure 11.2 is likely to be low (certainly lower than the one for the variables in Figure 11.1)

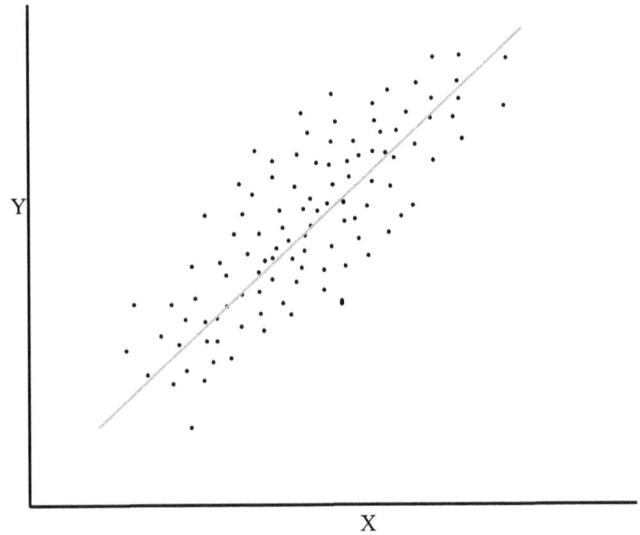

Figure 11.1 Linearity.

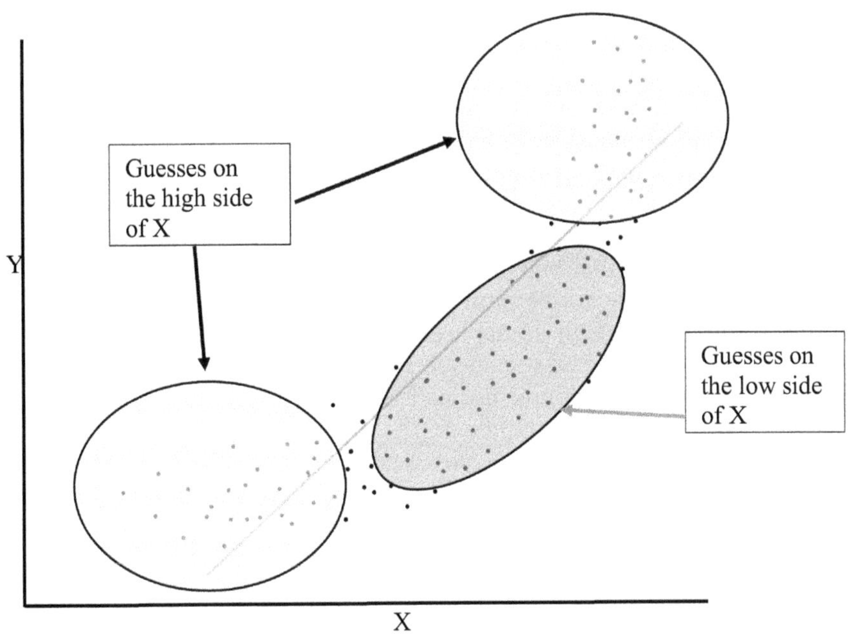

Figure 11.2 Nonlinearity.

despite the existence of a clear relationship between the two variables. The relationship is not a linear one, however.

Related to both independence and linearity is the idea of homoscedasticity. That's a word, right there. **Homoscedasticity** is the assumption that error terms are not correlated, and we can check specifically for this using residual plots (more on this in Chapter 12). Both violations of independence and of linearity can produce **heteroscedasticity**, or a patterning in error terms. The scatterplot seen in Figure 11.2 exhibits heteroscedasticity.

Level of Measurement

Clearly, the two variables should be *truly* metric, truly interval or ratio. In human sciences, we have a history of treating some ordinal variables as if they are metric. Pearson r has been used in circumstances like these, and without grievous calamity, but to do so is not encouraged. There are tests designed to work specifically with ordinal data (like Spearman ρ), and these are usually to be preferred in such circumstances.

Paired Scores

For Pearson r, for each case (for each row in the spreadsheet), there should be valid values for each X and Y variable (that is, for each of the two columns that we will use in the analyses). Often in criminological research, data are missing; sometimes measurements are incomplete, or subjects do not respond to all items. Valid values for both X and Y are needed for each case; if cases are not complete, they are deleted listwise (meaning that if information for either X or Y is missing for a case, that case can't be used in the analysis).

Distributional Normality

Ideally, both X and Y should be univariate normal (meaning that they are mostly symmetrical – not terribly skewed – and neither excessively leptokurtic nor platykurtic) and bivariate normal (meaning that a scatterplot of the X and Y variables also shows symmetry and normal kurtosis). Now, there is some disagreement regarding the importance of bivariate normality (e.g., Kachigan, 1986; Warner, 2008). Some have argued that the univariate normality of each variable is something of a minimum requirement for Pearson r to operate correctly, and that bivariate normality (that is, symmetric homoskedasticity) is a greater concern than the metric nature of the variables. Yet another concern regarding distributional normality, though, is sample size: a smaller sample is likely to have a greater degree of variance and a more volatile mean. It makes good sense, then, for the researcher planning to use Pearson r to examine normality by creating histograms for both X and Y, and then also creating an X/Y scatterplot.

What is rather clearer, though, is that extreme scores and outliers can gum up the works (such extreme scores in any abundance can create something that begins to resemble heteroskedasticity). We have seen already how outliers can jack up a mean; well, they can most surely jack up a Pearson r. Here's an example.

We have two hypothetical variables: X and Y, and the values can be seen in Table 11.2.

Univariate normality can be assessed at least at an intuitive level using histograms:

Just looking at the two variables' histograms, one might safely conclude that they are roughly normal. In fairness, Y (Figure 11.4) has a slight negative skew and X (Figure 11.3)

Table 11.2 X and Y Values for Hypothetical Example

X	Y
7	5
9	7
12	5
6	6
9	5
8	7
5	8
7	6
8	8
10	6
11	5
8	7
6	8
7	6
9	6
5	6
6	8
7	6
9	7
4	8
7	9
7	6
9	6
5	4
15	2
6	6
7	8
8	5

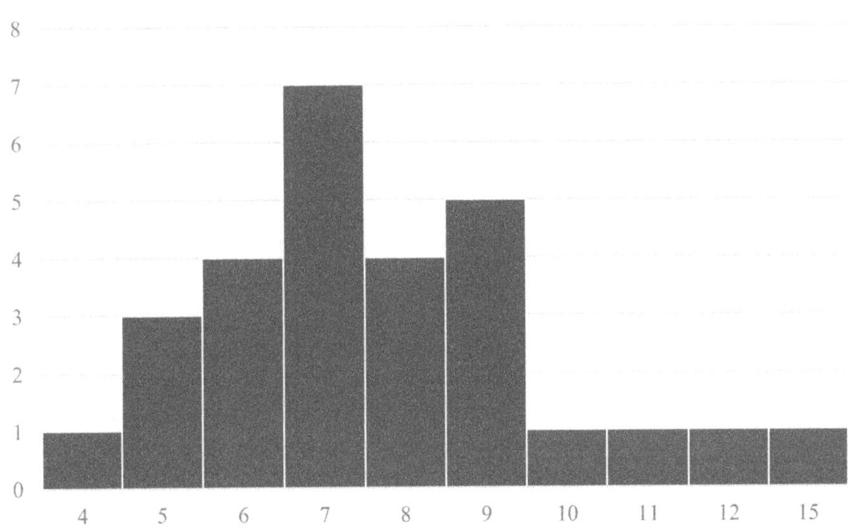

Figure 11.3 Histogram for X.

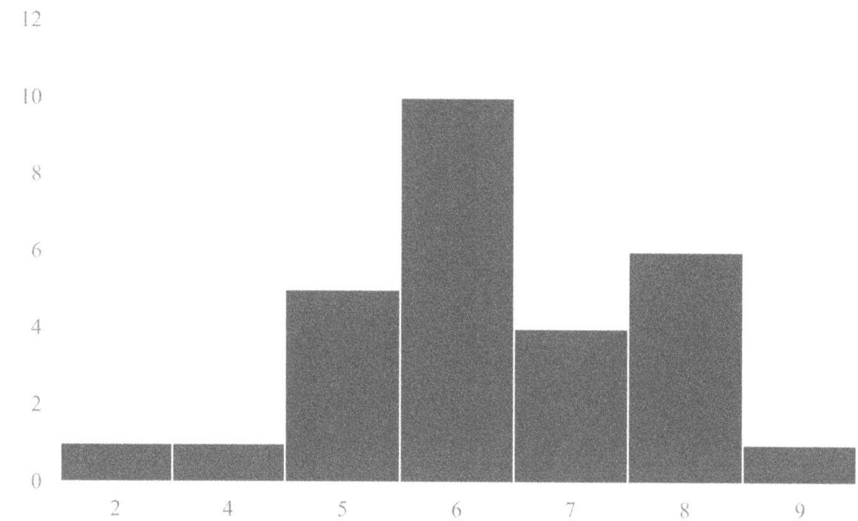

Figure 11.4 Histogram for Y.

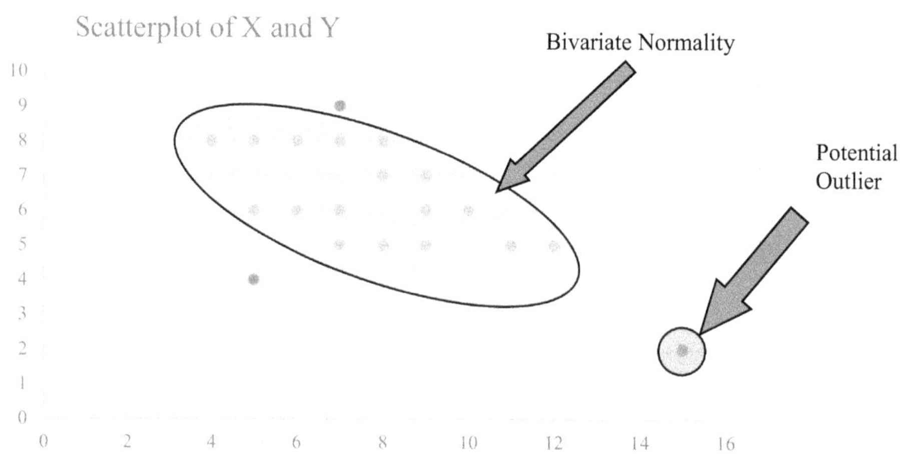

Figure 11.5 Scatterplot of X and Y, Bivariate Normality and Possible Outlier.

has a slight positive skew, but neither one looks overly problematic. It is when we look at a scatterplot of the two where we see something that might be an issue (see Figure 11.5).

As something of a standard, we tend to define outliers in terms of the number of standard deviations above or below a mean; under normal circumstances, a number that is greater or less than 3.3 standard deviations away from a mean is a *candidate* for identification as an outlier. In this circumstance, neither the 15 nor the 2 that correspond to our circled outcome would qualify as an outlier on their own, but as a point in the scatter, the distance from the group is evident. There are some tests we could perform to make a more definitive determination, but the purpose of the tests would be to determine whether or not we should remove the case from the analysis. My inclination in this case most likely would be to compare the values for r with and without the case in the mix, and if the impact of the case is large, removing it from the analysis (but not to delete it from the spreadsheet) becomes an option. Of course, all cleaning tasks like this would be detailed carefully in our analysis notes.

So, let's see what happens. If we calculate Pearson r for the range of scores as they came to us in Table 11.2, we derive a value of −0.5545 (we will talk about how to derive this shortly, as well as how to interpret it). If, however, we were to remove the one X/Y pair that might be an outlier (15, 2), we get −0.3220. Regardless of how these scores are interpreted, this change represents about a 40% change in outcome just from deleting one case. In essence, then, if that one case represents some kind of extreme, possibly misleading pair of outcomes, then its inclusion in the analysis would falsely (and meaningfully) inflate the apparent strength of the linear relationship between X and Y to a factor of 1.4. That the pair of outcomes happens to fall in line with the rest of the cases is reason to defend its inclusion, however; it may not be misleading if the values can be taken to be legitimate. If the analyst seeks a conservative approximation of the relationship between X and Y, though, listwise deletion is still a defensible choice.

Calculating Pearson's r

The calculation of Pearson's r is a ratio of two other terms: covariance (which we'll discuss in just a moment) and a measure of variance (really it is closer to a standard deviation without the sample size in the denominator). It makes sense to talk about them in turn.

There's a thing we call covariance. **Covariance** is the degree to which two numbers, well..., vary together (imagine that). As a numeric expression, it looks like this:

$$\text{Cov}_{x,y} = \frac{\Sigma(x-\bar{x})(y-\bar{y})}{n}$$

It is the sum of the products of paired differences between each score and its mean, divided by sample size. This isn't such a bad tool as a metric, but the bummer here is, as you may have concluded already, the "read-out" is arbitrarily large.

Covariance relates to the notion that scores for each of the two measures "move," or covary, together. If, for example, the value of X for one case is higher than the mean for X (or $\bar{X}$), is it also the case that Y for that case is also higher than *its* mean? We'll use the variables X and Y from just above. In order to talk about each score with any precision, we can begin by finding the variable's mean (say, the mean of X, $\bar{X}$) and then noting whether each individual score is above or below that mean (see Weisburd & Britt, 2007 for a similar treatment of this reasoning). The mean for X is 7.75. The mean for Y is about 6.286. For each score, then, we can note whether it is above or below the mean. I will note scores below the corresponding mean with a negative sign, and scores that are above the mean will be denoted with a positive sign (Table 11.3). But we will also want to find the degree to which X and Y for any one X/Y pair (that is, for each case) are similarly situated. For cases, like the first row, where both scores are below each mean, the product of signs (a negative times a negative) is positive; they're both on the same side of their corresponding means. The same is true for the second row, except both are *above* their means, not below. For the third row, though, x is above X's mean, but y is below Y's mean, so the product of signs is negative. Now as it turns out, the scores ($n = 28$) are evenly split in terms of their products of signs.

Table 11.3 X and Y Values for Hypothetical Example

X	$\bar{X} = 7.75$	Y	$\bar{Y} = 6.286$	*Product of Signs*
7	–	5	–	+
9	+	7	+	+
12	+	5	–	–
6	–	6	–	+
9	+	5	–	–
8	+	7	+	+
5	–	8	+	–
7	–	6	–	+
8	+	8	+	+
10	+	6	–	–
11	+	5	–	–
8	+	7	+	+
6	–	8	+	–
7	–	6	–	+
9	+	6	–	–
5	–	6	–	+
6	–	8	+	–
7	–	6	–	+
9	+	7	+	+
4	–	8	+	–
7	–	9	+	–
7	–	6	–	+
9	+	6	–	–
5	–	4	–	+
15	+	2	–	–
6	–	6	–	+
7	–	8	+	–
8	+	5	–	–

We can take things further, of course. Weisburd and Britt (2007) point to what we've just done as the basic reasoning that allows Pearson *r* to operate, but as we have more information available to us, let's put it to use. What we will do is to find the actual difference from the mean for each score, and then for each case, multiply the difference for *X* times the difference for *Y*, like in Table 11.4.

Much like other processes we've seen, we will then sum the products. In this case, the sum is –52.00. This measure of covariance will find its home in the numerator of the Pearson equation.

The calculation of "variance" remains and is in ways notably similar to (although not the same as) what we've done under the auspice of variance earlier. That measure of variance is the square root of the product of added differences from the mean for each score.

Pearson's *r*, then, is calculated:

$$r = \frac{\sum \left(X_i - \bar{X}\right)\left(Y_i - \bar{Y}\right)}{\sqrt{\left(\sum \left(X_i - \bar{X}\right)^2\right)\left(\sum \left(Y_i - \bar{Y}\right)^2\right)}}$$

Table 11.4 X and Y Values for Covariance in Hypothetical Example

X	$\bar{X} = 7.75$	Y	$\bar{Y} = 6.286$	*Product of Differences*
7	−0.75	5	−1.286	0.9645
9	1.25	7	0.714	0.8925
12	4.25	5	−1.286	−5.4655
6	−1.75	6	−0.286	0.5005
9	1.25	5	−1.286	−1.6075
8	0.25	7	0.714	0.1785
5	−2.75	8	1.714	−4.7135
7	−0.75	6	−0.286	0.2145
8	0.25	8	1.714	0.4285
10	2.25	6	−0.286	−0.6435
11	3.25	5	−1.286	−4.1795
8	0.25	7	0.714	0.1785
6	−1.75	8	1.714	−2.9995
7	−0.75	6	−0.286	0.2145
9	1.25	6	−0.286	−0.3575
5	−2.75	6	−0.286	0.7865
6	−1.75	8	1.714	−2.9995
7	−0.75	6	−0.286	0.2145
9	1.25	7	0.714	0.8925
4	−3.75	8	1.714	−6.4275
7	−0.75	9	2.714	−2.0355
7	−0.75	6	−0.286	0.2145
9	1.25	6	−0.286	−0.3575
5	−2.75	4	−2.286	6.2865
15	7.25	2	−4.286	−31.0735
6	−1.75	6	−0.286	0.5005
7	−0.75	8	1.714	−1.2855
8	0.25	5	−1.286	−0.3215

The numerator is our measure of covariance; it is the summed product of the $X - \bar{X}$ comparison for each individual case. Again, here we are denoting the degree to which each case's score on X in relation to $\bar{X}$ is similar to each case's score on Y relative to $\bar{Y}$. That covariance, added up across all scores, represents the total amount of variability that is attributable to within-case commonalities – shared variability. The measure of variance in the denominator is the amount of variance from the mean for X and Y separately.

Pearson r can be greatly simplified as:

$$r = \frac{\operatorname{cov}(X,Y)}{\sigma_X \sigma_Y}$$

thus, we can see the denominator's nature as the product of squared deviations from each variable's mean.

Let's finish our example. We've already found the numerator. We need the value for the denominator, and the process should be quite familiar by now (see Table 11.5).

Table 11.5 X and Y Values for "Variance" in Hypothetical Example

X	$(X - \bar{X})^2$	Y	$(Y - \bar{Y})^2$
7	0.5625	5	1.65306122
9	1.5625	7	0.51020408
12	18.0625	5	1.65306122
6	3.0625	6	0.08163265
9	1.5625	5	1.65306122
8	0.0625	7	0.51020408
5	7.5625	8	2.93877551
7	0.5625	6	0.08163265
8	0.0625	8	2.93877551
10	5.0625	6	0.08163265
11	10.5625	5	1.65306122
8	0.0625	7	0.51020408
6	3.0625	8	2.93877551
7	0.5625	6	0.08163265
9	1.5625	6	0.08163265
5	7.5625	6	0.08163265
6	3.0625	8	2.93877551
7	0.5625	6	0.08163265
9	1.5625	7	0.51020408
4	14.0625	8	2.93877551
7	0.5625	9	7.36734694
7	0.5625	6	0.08163265
9	1.5625	6	0.08163265
5	7.5625	4	5.2244898
15	52.5625	2	18.3673469
6	3.0625	6	0.08163265
7	0.5625	8	2.93877551
8	0.0625	5	1.65306122
Σ=	147.25	Σ=	59.7142857

Now we have all the bits and pieces we need.

$$r = \frac{-52.00}{\sqrt{(147.25)(59.7142857)}}$$

$$r = \frac{-52.00}{\sqrt{8{,}792.92857}}$$

$$r = \frac{-52.00}{93.7706168}$$

$$r = -0.554544715$$

...or about -0.55, just as we saw earlier when we included all of the cases in the example. Thus, we can say that there is a defensibly strong correlation between our hypothetical variables *X* and *Y*. What that means in fullness is left to you to determine.

There's Always Another Way...

Conceptually, yet another way of approaching r is this (e.g., Warner, 2008):

$$r = \frac{\Sigma z_x z_y}{n-1}$$

Seen this way, it is an average of a kind – specifically, an averaging of paired scores. One of the coolest things about this, and more, about this particular way of understanding Pearson r, is that because it uses z scores, the scales of the variables we use do not have to be similar; scale is immaterial, which is pretty cool just generally.

It bears noting that at least one authoritative source (Warner, 2008) does not present the first, main definitional formula for Pearson r and instead only discusses the formula directly above; the conceptual simplicity of the standardized version is appealing, to be sure.

Let's work on one. Table 11.6 below presents our two variables.

Our null hypothesis is:

$$H_0: \rho_{xy} = 0$$

Or that r is zero, meaning that there is no relationship between the two variables. The ρ you see there is intended to denote the correlation as it might exist in the population (because that is what we are trying to determine). In words, it is saying that the null hypothesis is that there is no relationship between X and Y in the population (as revealed by the sample before us), and that the best predictor of Y in the population is the mean of Y. The alternative hypothesis, then, would be:

$$H_A: \rho_{xy} \neq 0$$

Notice that the variables in Table 11.6 do not operate in similar ranges. Again, that matters not at all. First, we need to find the means and standard deviations in order to create our z scores. The mean for X is 37.9 and the mean for Y is 0.783. If we do the rest of the standard deviation calculation, we get a standard deviation of 6.285 for X and 0.376 for Y (Table 11.7).

$$r = \frac{\Sigma z_x z_y}{n-1}$$

$$r = \frac{4.772}{6-1}$$

Table 11.6 Variables for Pearson *r*

X	Y
34.1	0.4
48.5	1.3
40.2	1.0
36.7	0.8
29.9	0.3
38.0	0.9

Table 11.7 Calculation of Pearson r

X	Y	$z_x = \frac{x - \bar{x}}{s_x}$	$z_y = \frac{y - \bar{y}}{s_y}$	zxzy
34.1	0.4	−0.6046	−1.0185	0.6157
48.5	1.3	1.6864	1.3727	2.3149
40.2	1.0	0.3659	0.5756	0.2106
36.7	0.8	−0.1909	0.0443	−0.0085
29.9	0.3	−1.2728	−1.2841	1.6344
38.0	0.9	0.0159	0.3010	0.0049
Σ:				4.772

$$r = \frac{4.772}{5}$$

$$r = 0.9544$$

That is an *exceedingly* high correlation, indicating that our hypothetical X and Y variables are *very* closely related. In other words, as X varies, it would appear that Y also varies. Now, this is not the same thing as saying that Y varies *as a function* of X unless very specific laboratory conditions are present for these data (conditions, in fact, that would be so extraordinarily burdensome as to make their achievement nearly impossible in a criminological context): the inference of cause from correlation alone is not defensible.

If we look at a scatterplot of the two hypothetical variables, we can see this rather strong relationship at work:

Weisburd and Britt (2007) tell us that in human sciences, we can interpret r values around 0.1 (or −0.1) as "low," scores around 0.3 (or −0.3) as "moderate" and anything greater than 0.5 (or −0.5) as "large" (see also Cohen, 1988).

Significance of r

The determination of statistical significance for Pearson r is accomplished using a t distribution according to the formula:

$$t = \frac{r - \rho_0}{\sigma_{\bar{x}_r}}$$

Where ρ_0 (again, the Greek character *rho* is used to denote a population parameter corresponding to r) will be 0 (because that's the null hypothesis), and the standard error of r ($\sigma_{\bar{x}_r}$) is:

$$\sigma_{\bar{x}_r} = \frac{\sqrt{1 - r^2}}{\sqrt{n - 2}}$$

With some mathematical reasoning that we won't present or explain here (see, e.g., Warner, 2008), our p value for r can be derived:

$$t = \frac{r^2 \times \sqrt{n - 2}}{\sqrt{1 - r^2}}$$

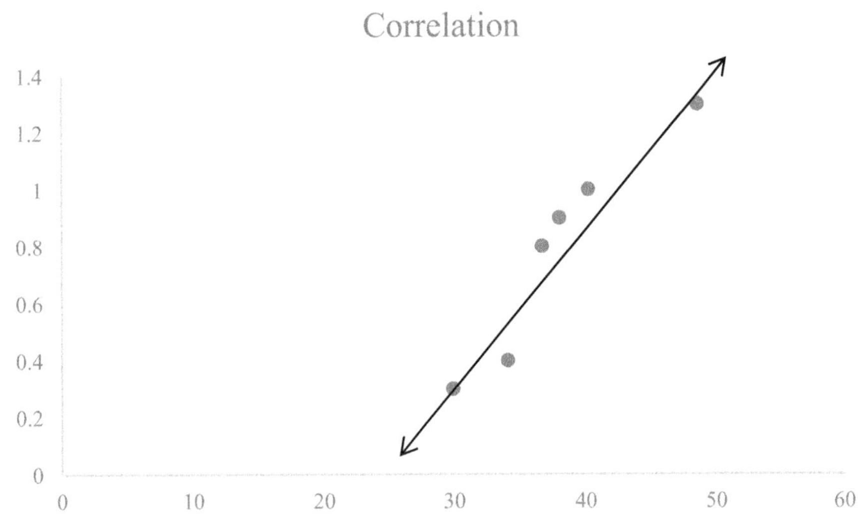

Figure 11.6 Scatterplot of Hypothetical Example of Correlation with Trend Line.

I think it also bears observing here that the formula for significance just above has a term for sample size in it, as well as a degrees-of-freedom correction (that "minus-two" business). Sample size is important for t, of course, as t varies according to sample size (to a limit). Our example here uses $n=6$, and from a Central Limit Theorem perspective, we would want there to be a "large" (and truly EPSEM random) sample (with replacement), that is $n=30$ or greater, for reliably stable results in support of inference. As a result, although the example operates here and can be utilized in a local sense, if the result is to be used in support of inference regarding a population, the larger the sample, the better, and our example here would be too small to support sound inference. I used a small sample here for illustration purposes.

Let's put it to use for the previous example.

$$t = \frac{0.9544^2 \times \sqrt{6-2}}{\sqrt{1-0.9544^2}}$$

The value for r we found above already, and n is the number of cases. Let's continue.

$$t = \frac{0.9109 \times 2}{\sqrt{1-0.9109}}$$

$$t = \frac{0.9109 \times 2}{\sqrt{0.8091}}$$

$$t = \frac{1.822}{0.2985}$$

$$t = 6.10.$$

Again, the null hypothesis is that $r_{xy}=0$. The alternative hypothesis, then, is that $r_{xy} \neq 0$. Thus, we seek a two-tailed test (and that should almost always be the way things are for correlations). Our t_{crit} for a two-tailed test at $p=0.05$ ($n-1=$ 5 degrees of freedom) is 2.571

(see Appendix B). Our t_{obs}, seen above, is 6.10. Because the t_{obs} falls well out into the rejection region on the t curve at 5 degrees of freedom, we can reject the null hypothesis. The correlation between X and Y in our hypothetical example is statistically significant, again, with the rather important sample size caveat above.

Nota Bene: Curves Ahead

Pearson r is for circumstances where we seek to speak to the linear relationship between two truly continuous or metric variables. Pearson r is a powerful technique, but unlike ANOVA and t test, it can be somewhat touchy with regard to the violation of assumptions. The sensitivity of Pearson r comes at something of a cost in terms of robustness.

Remember how I said that Pearson r is really a test of the linearity of the relationship between two variables? Well, there are places in the many spheres of human inquiry where we might find linear relationships between metrically measurable phenomena. In human sciences, though, linear relations are both seemingly more rare and less intriguing than those that have some... flavor. What is more, many things that really interest us are not meaningfully metric in nature. We will need, then, to have analytical tools that can accommodate nonlinear phenomena or can operate over rougher, non-metric terrain. The next section will present techniques that offer capacities not present with Pearson r.

Other Kinds of Correlation Analyses

Although they are beyond the scope of this introductory text, there are techniques for nonlinear correlations analyses. Székely and colleagues (2007) developed a distance-based technique for testing the relationship between two variables that is not fettered by the notion of linearity. Wang and colleagues (2015) have created a measure called continuous analysis of variance, or CANOVA, that also does not depend upon linearity.

For our purposes, though, we shall discuss a few techniques that are in more common usage and are likely to be found in most statistical software.

Spearman ρ

When we have ordinal data, it is preferable to use tests designed specifically for ordinal/ordinal relations. Spearman ρ (rho) is just such a test. Spearman ρ operates to measure the **monotonic** relationship between two ordinal variables. The variables should be monotonic (meaning that they are always decreasing or always increasing across their ranges, even when the relationship is nonlinear) because the technique uses rank ordering to introduce a component of quantitative meaningfulness to the affair. We can also use Spearman ρ when there are issues with the variables such that normality cannot be assumed.

Basically, what we do is rank each score for both variables and then perform a correlation analysis on the ranks. In cases of ties, one could either use a more complicated formula (actually, it would be the bigger form of Pearson above, but with ranks instead of scores), or we could do this:

$$\rho_s = 1 - \frac{6\Sigma\Delta^2}{n\left(n^2 - 1\right)}$$

Simply rank each score (assigning an averaged rank for ties), subtracting the rank for X from the rank for Y, then squaring the difference (we used the Greek Δ to represent the difference between ranks), adding it up, then multiplying the sum by 6, and then dividing by the indicated denominator. Serve immediately.

As it turns out, there are several tests, like the Wilcoxon Rank Sum test (an alternative to t test), also called Mann–Whitney U, and all use a similar ranking logic. They allow for a comparison of the independence of two groups by ranking scores.

Point Biserial, r_{pb}

As we discussed in the chapter on t tests, point biserial is a technique that allows us to determine the relationship between a metric variable and a *true dichotomous* variable (other techniques are preferred over "dichotomizing" or dividing a continuous or ordinal variable into two groups). We can treat point biserial as a measure of effect size (as we did earlier) or as a measure of the strength of relationship between group status and the metric dependent variable. Point biserial and t test are closely related. Warner (2007) illustrates how r_{pb} and t can be exchanged one for the other:

$$t = \frac{r_{pb}\sqrt{n-2}}{\sqrt{1-r_{pb}^2}}$$

is related to

$$r_{pb} = \sqrt{\frac{t^2}{t^2 + df}}$$

To derive a point biserial correlation coefficient, we use

$$r_{pb} = \frac{\bar{x}_a - \bar{x}_b}{s_{n-1}}\sqrt{\frac{n_a n_b}{n^2}}$$

with

$$s_{n-1} = \sqrt{\frac{1}{n}\Sigma(x-\bar{x})^2}.$$

The $\bar{x}$ terms above are means for the two groups created by the dichotomous variable, and s_{n-1} is a form of standard deviation. Again, we can see a signal-and-noise shape to this measurement, just like what was present with t test and ANOVA.

Phi (ϕ)

If we have a circumstance where we want to know about the strength of the relationship between two dichotomous variables, we would use phi (ϕ). We will use ϕ in Chapter 13 which presents nonparametric techniques like chi square (represented as χ^2), as it is commonly used in 2 × 2 applications of chi square (which will make more sense very soon). Phi, like Spearman ρ and point biserial, is a special case of Pearson r (e.g., Warner, 2007).

Table 11.8 Cell Designations for the Calculation of Phi (ϕ)

		Values of one of the variables	
		Value 1	*Value 2*
Values of the other variable	Value 1	*a*	*b*
	Value 2	*c*	*d*

If we have a chi-square (χ^2) value (which we have not yet discussed, but will treat in detail in Chapter 13), we can use it to derive a measure for phi:

$$\phi = \sqrt{\frac{\chi^2}{n}}$$

If we don't have χ^2, we can derive ϕ:

$$\phi = \frac{bc - ad}{\sqrt{(a+b)\times(a+c)\times(b+d)\times(c+d)}}$$

where the letters correspond to frequency values in each cell of a 2 × 2 table. Table 11.8 illustrates the letters.

Summary

With *t* tests and ANOVA, the two variables included in the analyses don't truly meet like equals. The metric dependent variable is really the star of the show; even if group differences are interesting, they're interesting in terms of the dependent variable. Correlation analyses operate a little differently in that they seek to determine the degree to which two variables move together. Pearson *r* is something of a gold standard technique, useful for metric variables. Understanding and paying attention to assumptions of the test matters greatly; Pearson *r* may not be as robust as ANOVA against violation of test assumptions. Where the two variables are not metric, we have other techniques. Spearman ρ uses rank orders of variable values to derive a very similar comparison of co-variation between two variables, and point biserial and phi both work in their respective bandwidths to provide hypothesis testing in settings where variables are categorical or differently ordered.

Main Take-Aways

- Pearson *r* and Spearman ρ both treat the two variables as equals in the analyses.
- Paying close attention to test assumptions might mean choosing a different test from the one you had planned to use. Tests like Spearman ρ and point biserial should not be overlooked when considering correlation-related hypotheses.
- Pearson *r* is a test of linearity; if two variables covary but in a nonlinear way, Pearson *r* will not operate optimally, and results will give an incomplete picture of the relationship.

Key Terms (in order of appearance)

correlation matrix
linearity
levels of measurement
covariance
monotonic
Pearson r
homoskedasticity
paired scores
significance of r
point biserial
independence
heteroskedasticity
distributional normality
Spearman rho (ρ)
phi (Φ)

Chapter 12

An Introduction to Bivariate Linear Regression

Much of what has come before this point uses the same basic bits of reasoning: sum of squares divided by degrees of freedom, signal-to-noise ratios, and relative likelihood to sampling distributions that stand as proxies for the unknown values of populations. What we will discuss in this chapter is more of the same reasoning but taken further; regression analysis is merely a bigger vessel built on the plans of the dinghies and yachts we have contemplated thus far.

This chapter offers a conceptual introduction and some practical utilization of the basic ideas of regression. Quite literally, there are throngs of volumes written and entire courses taught on regression alone, and this chapter cannot hope, and shall not try, to consolidate it all into one place. My aim here is to provide the framework, interest, and connective fertility for you so that when you come to engage with that literature or with regression applications beyond our compass here, you are comfortably conversant.

Now, it should be observed that regression has been used for decades in all manner of scientific settings. Chemists use it, zoologists use it, sociologists use it, psychologists use it.... Its impact on how scientists think about the things they study is truly outsized, just as we have discussed with regard to correlation, but even to a greater extent. Regression is so deeply intuitive and directly relatable to the way scientists tend to ask questions that it is fair to ask if its intuitive nature (and thus some of its logical bugs) are cause or effect of how scientists think. I tend to think that it is both cause and effect in actuality; it arose to solve a problem, but it did so in such a compelling and narrative-supportive way, scientists began to shape their questions to fit the model regression presents.

Introduction to Bivariate Linear Regression

Regression and correlation, and when it comes to it, regression and ANOVA, as we have discussed them are all closely related bits of reasoning. The relationship goes well beyond the whole Central Limit Theorem inference business, too, but in many ways, it is a function of that logic. ANOVA, Pearson correlation and regression do related things in similar but distinct ways. All of them use something of a signal-to-noise ratio operability but operationalize signal and noise in different ways. For *t* test, a pooled or averaged standard error does good service as our measure of noise. For ANOVA, because the amount of variance involved is larger (more levels mean more complexity), we need to work the noise measure a little differently. Sure, we will pool standard error in a way, but for ANOVA, we need to measure that error according to a different, even case-wise comparison with group means. Because of the greater complexity in ANOVA, greater efficiency is necessary for the creation

DOI: 10.4324/9781032638591-12

of measures. We need to parse the total pile of error into groups so that we can offer a sound and more efficient comparison of signal and noise. But as we move into greater efficiency through the partitioning of sums of squares, we also open the possibility of using yet other strategies to determine patterning at work in our data. Regression represents a further refinement of what we've done so far, and in actuality, it adds only a few new bits of reasoning to do so.

This humble set of refinements allows for something of great potential value that statisticians often are asked to provide: **prediction**. What will the demand for jail beds look like? How many officers will need to be trained and certified by next year to meet staffing and service needs? Regression can help with these and similar questions.

Kids of a certain age are wont to make arguments to parents and teachers about the utility of the math they're forced to take while in grammar and high school. Well, what is about to happen here might be a little embarrassing for some folks who have made that argument themselves. The logic that makes all this possible is something you've already handled, probably in middle school or even before. Heck, it might well be the very example to which you gestured in frustration, saying, "Now when am I going to need to be able to do *this* as an adult?!" Yep, that time has come.

Do you recall this?

$$y = m(x) + b$$

It is the mathematical equation that iterates the drawing of a straight line on a coordinate grid. The term b corresponds to the point of the Y intercept, and m is an expression of slope for every unit of x. Well, that's the beating heart of regression. Typically, though, statisticians write the same formula this way:

$$y = a + b(x) + \varepsilon$$

Such that a is the **Y intercept** and **b** becomes that slope value (or weight), x is the value of our independent (or predictor) variable, and ε represents error, or the **residual**. In Figure 12.1, we can see the constituent parts of the simple linear equation that is at the center of regression modeling. The b term, or slope, is derived:

$$b = \frac{\text{change in } y}{\text{change in } x}$$

Say we have the following data:

Nine cases, each with a value corresponding to variable X and variable Y are presented in Table 12.1. Figure 12.2 shows the same data in graphic form. Remember to keep an eye on the scaling – the two axes use different scales. The change in Y looks smaller than the change in X, but that difference is because the major gridlines for Y are by units of 5, whereas the major gridlines for X are by ones.

For every unit increase in X, there is a two-unit increase in Y. The Y intercept is 4. If the slope is the change in Y over the change in X, then that means $b = 2/1 = 2.0$. Therefore, our regression equation would be:

$$y = 4 + 2x$$

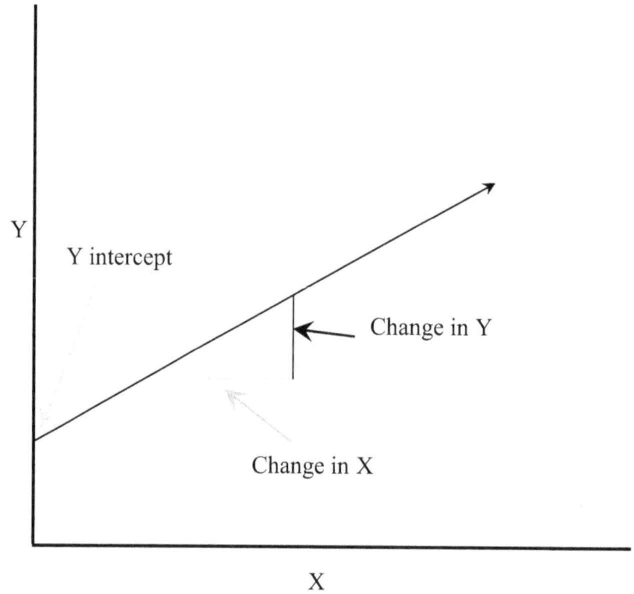

Figure 12.1 Basic Logic of a Linear Equation.

Table 12.1 Hypothetical Data for the Logic of Regression

X	Y
4	12
7	18
6	16
9	22
5	14
6	16
8	20
5	14
7	18

This expression, then, ends up being a rather efficient way of explaining the relationship between X and Y. Recall before how we talked about the pursuit of being able to speak in terms of Y as a function of X ($y=f(x)$)? Well, this ends up quite literally giving us precisely that capacity. Y is 2 times X (plus 4). Pretty slick, yeah?

Well, there's a catch. There's almost always a catch. Human data don't often fall on a perfectly straight line. Heck, I can't *skate* a straight line, and ice skates, by definition, *are* straight lines. Data are often scattered affairs, with perhaps something of a trend or pattern, but almost never lining up like ants marching home from a picnic. What we need is some way to speak to variance of Y from some expected value across values of X.

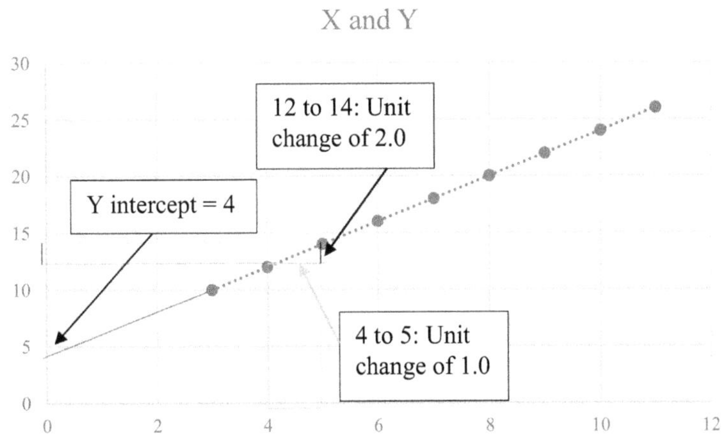

Figure 12.2 Example of the Logic of Regression.

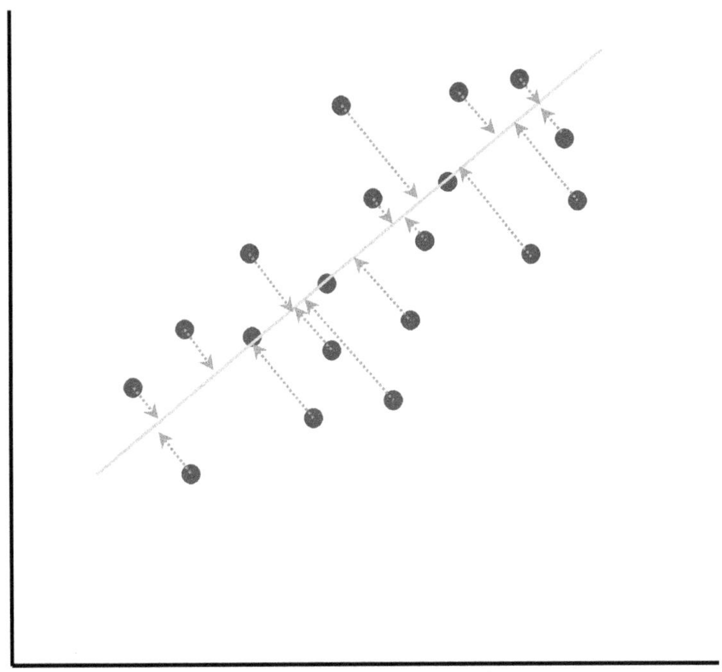

Figure 12.3 Linear Regression and "Error."

Hmm.

Most of the tools we've used thus far set us up nicely for intuiting how to do precisely this. We need a mean of a kind, and then we need to compare each outcome to that mean, just as we did with standard deviation and Pearson *r*.

The best guess of a value of a single score, if we don't know the actual score, is the mean of that variable (see the gray horizontal line in Figure 12.4). In lieu of other

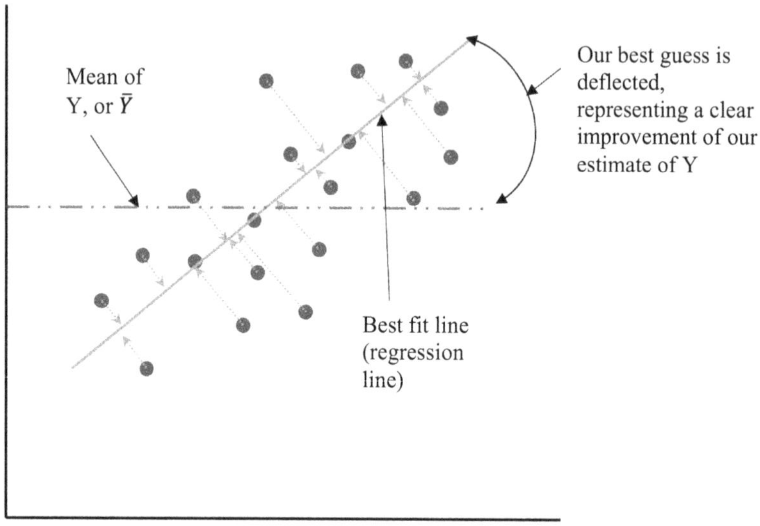

Figure 12.4 Regression Line and Mean as Estimates.

information about a particular case, the mean of the variable is our best guess at that score. If we know the mean height of humans living today, we could use that mean as a guess for your height. That it isn't precisely correct isn't the point, of course. The line in Figures 12.1–12.4 all have a slope. We'd probably be wrong with our mean-inspired guess, though, but the degree of wrongness itself is useful. If two variables are related, say height and weight, we might be able to use one variable to improve the quality of the guesses we make.

What we can do in order to maximize the efficiency of our guessing to is to fit a line to our points such that the distance between each point and our line (like our mean or... well, something else) is the smallest amount possible (see Figure 12.4). Instead of iteratively finding the line of maximal fit, that perfect point of smallest total error, though, we can use something we've already used to find it straight away.

Now, to get a precise **slope** for our maximum-fit, minimum-error line, we need some mathematical strategy that is a little more sophisticated than the "change in *Y*/change in *X*" thing we used at the beginning of the chapter. Recall that covariance/variance ratio we used with Pearson *r*? Yep, it is back, or rather, something much like it:

$$b = \frac{\Sigma(x - \bar{x})(y - \bar{y})}{\Sigma(x - \bar{x})^2}$$

This is yet another example of finding a **sum of squared errors** and comparing them as we do in ANOVA. The above formula is essentially $\delta y/\delta x$ (or the change in *y* divided by the change in *x*, just as we had before). Once we have *b*, we can derive the Y intercept:

$$a = \bar{y} - b\bar{x}$$

An alternative method for deriving the slope of our regression line is:

$$b = r\left(\frac{\sigma_y}{\sigma_x}\right)$$

where r is the Pearson correlation between X and Y, and the σ-terms are the standard deviations corresponding to each variable. This formula shows rather clearly how closely related regression and Pearson correlation are.

Let's try an example.

Example of Bivariate Regression

In Table 12.2, we have three columns: the first is the officer's made-up badge number. Next in the table is a column for age and then the last is a column for the number of years the officer has been a police officer. We want to see the degree to which there is a relationship between officer age and officer years on the job. The null hypothesis is that there is no relationship between the two variables, that the value for b will be 0, or that the best guess for any value of Y is the mean of y, or $\bar{y}$. Somewhat weirdly, our independent variable will be age; our model will determine the degree to which variance in years on the job is accounted for by variance in age.

Our task, then, is to derive a regression line using our $\delta y/\delta x$ formula.

$$b = \frac{\Sigma(x-\bar{x})(y-\bar{y})}{\Sigma(x-\bar{x})^2}$$

The mean for age (our X) is 44.533, and the mean for years of service (our Y) is 16.467. We will need a table.

Table 12.3 will suffice nicely.

Table 12.2 Officer Age and Years on the Job

Officer	*Age (X)*	*Years (Y)*
1	37	15
2	50	9
3	37	11
4	49	24
5	41	22
6	57	33
7	50	8
8	43	19
9	39	15
10	25	3
11	47	15
12	59	29
13	36	12
14	49	18
15	49	14

Table 12.3 Calculations for Slope for Regression Example

Officer	*Age (X)*	*Years (Y)*	$(x-\bar{x})$	$(y-\bar{y})$	$(x-\bar{x})(y-\bar{y})$	$\Sigma(x-\bar{x})^2$
1	37	15	−7.5333	−1.4667	11.0489	56.7511
2	50	9	5.4667	−7.4667	−40.8178	29.8844
3	37	11	−7.5333	−5.4667	41.1822	56.7511
4	49	24	4.4667	7.5333	33.6489	19.9511
5	41	22	−3.5333	5.5333	−19.5511	12.4844
6	57	33	12.4667	16.5333	206.1156	155.4178
7	50	8	5.4667	−8.4667	−46.2844	29.8844
8	43	19	−1.5333	2.5333	−3.8844	2.3511
9	39	15	−5.5333	−1.4667	8.1156	30.6178
10	25	3	−19.5333	−13.4667	263.0489	381.5511
11	47	15	2.4667	−1.4667	−3.6178	6.0844
12	59	29	14.4667	12.5333	181.3156	209.2844
13	36	12	−8.5333	−4.4667	38.1156	72.8178
14	49	18	4.4667	1.5333	6.8489	19.9511
15	49	14	4.4667	−2.4667	−11.0178	19.9511

If we calculate the sums for the last two columns, we get 664.267 and 1103.733, respectively. Let's plug.

$$b = \frac{664.267}{1103.733}$$

$$b = 0.602$$

Or thereabouts (I rounded at the end in order to make a later point). Now we need the Y intercept.

$$a = \bar{y} - b\bar{x}$$

$$a = 16.467 - 0.602(44.533)$$

$$a = 16.467 - 0.602(44.533)$$

$$a = 16.467 - 26.809$$

$$a = -10.342$$

This means our regression equation would be:

$$y = -10.342 + 0.602x$$

Take a look at what Excel gives us (Figure 12.5):

Excel inserts its own trendline, complete with regression equation. You will notice, I imagine, that there are slight differences in the values Excel got from the values we got. Those differences are the result of rounding error. I rounded early in the process, but just a little.

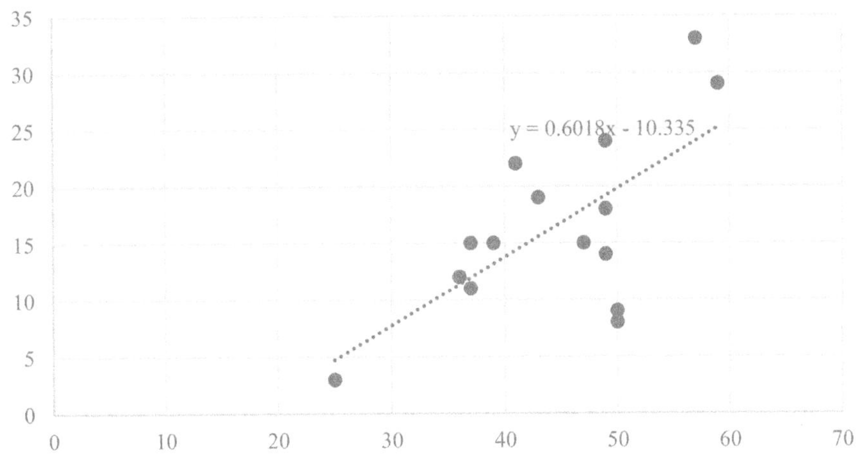

Figure 12.5 Regression, Age by Years for Police Officers.

In Excel, the line was derived on the basis of precision out to several decimal places. I only used four in our calculations. The point made here is that early rounding can produce a propagation of error such that outcomes can be notably different. The two results aren't all *that* different, of course, but the point is made that differences can propagate when we round early.

So what? Rather, now what? We have the line, and indeed Excel could have done this for us without all that figuring. Two questions arise from what we've done so far. First, we want to know about the likelihood of our results being a function of chance (that is to say, are our results statistically significant), and we also want to know something about the strength of the relationship, or the overall quality of our model.

The first thing, the significance bit, we do in two steps. First, we need to introduce a particular functionality of the equation we made just above. For each *X*/*Y* pair of points, we can use the expression to create a predicted Y (or $\hat{y}$ – I put on that little hat to designate a prediction) value. Let's take a look:

If years on the job was solely a function of the equation (a combination of the *Y* intercept and age), the values in the last column in Table 12.4 would be spot on (meaning that they would be vanishingly small). As it happens, there is a difference between actual and predicted scores, just like what we have when we do standard deviations (or standard errors, as is the case here).

So, we need an estimate of the standard error for the denominator of our comparison, and then we can perform what is, in essence, a single-sample *t* test. First, the standard error:

$$\sigma_{\bar{x}b} = \sqrt{\frac{\Sigma(y-\hat{y})^2/(n-2)}{\Sigma(x-\bar{x})^2}}$$

Where $\hat{y}$ represents the predicted value of y for each point (we can use our equation to produce a predicted value for each *X*/*Y* pair). We have already derived the $(y-\hat{y})^2$ business

Table 12.4 Predicted Values of Y ($\hat{y}$), $(y-\hat{y})$ and $(y-\hat{y})^2$

Officer	*Age (X)*	*Years (Y)*	$\hat{y}$ (or predicted y): $\hat{y} = -10.342 + 0.602x$	$(y-\hat{y})$	$(y-\hat{y})^2$
1	37	15	11.932	3.068	9.413
2	50	9	19.758	−10.758	115.735
3	37	11	11.932	−0.932	0.869
4	49	24	19.156	4.844	23.464
5	41	22	14.340	7.660	58.676
6	57	33	23.972	9.028	81.505
7	50	8	19.758	−11.758	138.251
8	43	19	15.544	3.456	11.944
9	39	15	13.136	1.864	3.474
10	25	3	4.708	−1.708	2.917
11	47	15	17.952	−2.952	8.714
12	59	29	25.176	3.824	14.623
13	36	12	11.330	0.670	0.449
14	49	18	19.156	−1.156	1.336
15	49	14	19.156	−5.156	26.584

in Table 12.4, and the $(x-\bar{x})^2$ bit from Table 12.3. We need to add the values, summing the squared errors for the two columns, and then all that's left is to plug-and-chug. $\Sigma(y-\hat{y})^2 = 497.954$ and $\Sigma(x-\bar{x})^2 = 1103.733$

$$\sigma_{\bar{x}b} = \sqrt{\frac{\Sigma(y-\hat{y})^2/(n-2)}{\Sigma(x-\bar{x})^2}}$$

$$\sigma_{\bar{x}b} = \sqrt{\frac{497.954/13}{1103.733}}$$

$$\sigma_{\bar{x}b} = \sqrt{\frac{38.304}{1103.733}}$$

$$\sigma_{\bar{x}b} = \sqrt{0.0347}$$

$$\sigma_{\bar{x}b} = 0.186$$

All that's left, then, is to do the t test, and that looks like this:

$$t = \frac{b-0}{\sigma_{\bar{x}b}}$$

The reason we subtract 0 from b is that we are testing the null hypothesis that b=0 (that there is no effect). This part is easy.

$$t = \frac{0.602-0}{0.186}$$

$$t = \frac{0.602}{0.186}$$

$$t = 3.237$$

Our t_{obs} is 3.237. Degrees of freedom for our t test here is $n-2$. So, for 13 degrees of freedom (it will always be a two-tailed test here) and with a p = value of 0.05, the t_{crit} value is 2.160, which means that once again, we can reject the null hypothesis.

A Few Final Notes

As we have said, there is much to discuss with regression, and much of this is well beyond the scope of this text to cover in detail. I would like to bring your attention to a few slick features about regression, though, that I hope will pique your appetite for more.

First, although we have been performing bivariate regression here, with some minor modifications, we can ratchet up the game and add multiple variables to our equation:

$$y = a + b_1x_1 + b_2x_2 + b_3x_3 + \cdots + b_nx_n + \varepsilon$$

We might want to know about the effects of several variables taken together on a single dependent variable; the independent terms taken together are called a variate and can be treated as a whole or in terms of the relative influence of the constituent terms.

Another point that bears our attention here is the idea of **explained variance.** If you recall from our conversation about Pearson r, we found out that we could talk about the degree to which one variable explained, or accounted for, variance in another. Well, one of the banner capacities of regression that we will not discuss here in detail is the expression r^2: *simply squaring the Pearson correlation* gives us an expression between 0 and 1 that articulates the degree to which unit changes in X accounts for unit changes in Y. This term is so pervasive and influential that other more sophisticated techniques (like logistic regression) have been modified to produce a "pseudo r^2" for the sake of interpretability (and, presumably, comparisons with other regression models).

One more cool thing about regression to which I alluded briefly near the beginning of the chapter was its capacity to predict things. The linear regression model can be drawn out past its instant limits; if our linear function is sufficiently sound (meaning that it is linear in nature, that it has a decent degree of fit, and that error terms are even across the range of the dependent variable), under the right circumstances that line can tell us things about values of Y, varying as a function of X (or our variate), even for conditions beyond those evident in our data.

Summary

The tools used up until now can be arranged in such a way as to make a statistical technique capable of offering insight into the nature of a relationship between an independent variable and a dependent variable. We are able to determine the degree to which a dependent variable changes under the influence of the independent variable. What is more, given a defensibly sound analysis, we are able to make predictions about values for our dependent variable based on the regression model. Like Pearson r, a linear relationship is assumed, which represents a kind of weakness in the technique, but even given the linear assumption, bivariate (and also multivariate) regression enjoys a kind of primacy among statistical techniques, even to the point where some other analytical techniques are given trappings that resemble regression's output.

Main Take-Aways:

- Linear regression, whether bivariate or multivariate, closely resembles a mathematical concept often taught in middle school.
- Slope, or the change in *Y*/change in *X*, gives us a crucial component in the regression equation.
- Regression allows for the additional capacity to predict values of *Y* on the basis of values of *X* (through the regression equation).

Key Terms (in order of appearance)

prediction	Y intercept	b
residual	slope	sum of squared errors
explained variance	r^2	

Chapter 13

Nonparametric Tests of Relation and Measures of Association

In a now-classic essay, Herbert Costner (1965) begins by gesturing grandly to a king's ransom of statistical measures of association; we are, he tells us, spoiled for choice. But choice, though, is a problem: how are we to know when to use any of the many measures of association at our disposal? Not infrequently, he tells us, many researchers balk at immersion into wrangling over which test is most appropriate for any given empirical setting. It is funny how little things have changed during the intervening decades. We have more choices now than ever before, but it is by no means a settled issue whether we are wealthier for the riches.

Costner also observes something vital that you and I have discussed already: in the human sciences, true metric measures are neither terribly plentiful, nor do they often deliver on their promises even when they can be tracked down in the wild. Arguments like Wilson's (1971), that "there seems little doubt that mathematically formulated techniques [that is, using interval- and ratio-level variables in multivariate analytical techniques like path analysis] will be used increasingly in substantive sociological research" (Wilson, 1971: 441) today seem not to have aged terribly well, at least to me. Although in fairness, Wilson was hardly mistaken in the short term, the results of that focus on mathematically formulated techniques simply have not borne abundant fruit (see, e.g., Weisburd & Piquero, 2008; Young, 2011; Ziliak & McCloskey, 2011), even in the era of "big data." *Indeed, in terms of the human sciences guided by the standard of the CLT and statistical significance alone, quantitative evidence suggests that quantitative evidence doesn't suggest much.*

And yet here we are, me writing and you reading ... *a statistics text.* Please do not take to the notion that just because Wilson's (1971) argument is dramatically overstated and staggeringly myopic, that because not a few commentators have observed that quantitative epistemic approaches to criminology have delivered disappointingly, even cruelly, that therefore we should abandon statistical technologies altogether for reasoning out answers to our human sciences questions. To do so would be to make the same mistake that Wilson made, only in the opposite direction. It seems clear that Costner understood these points before more than a few others did, and his understanding led him to expand on the thinking of Guttman (1941) and Goodman and Kruskal (1954) in order to seek measures of association that were not only capable of using variables with nominal or ordinal levels of measure, but that also spoke to the proportionate reduction of error (which we will discuss below).

The use of statistical technologies is *absolutely* worthwhile when studied phenomena in the world are revealable via statistical technologies. That sounds like a goofy thing to say, but it isn't: if epistemological strategies are not in consonance with the fullness of investigated human truth (e.g., DeValve, 2020b), calamity is likely to ensue. Ziliak

DOI: 10.4324/9781032638591-13

and McCloskey, like the statistical giants they cite, want us to attend to actual size of phenomena (as opposed to statistical significance), and to real error (as opposed to hypothetical standard error). Fair enough. The problem remains that such ideas as the real size of a criminological phenomenon (e.g., like strain) or the existence of real error are not ready fits to many criminological questions. Many of the things criminologists study are denizens of human interiority and, as such, they altogether defy measures like the ones Ziliak and McCloskey crave. For the same reason, though, we must be rather more cautious than we tend to be today in the utilization of statistical findings in criminological research.

We must also be more circumspect regarding the nature of the questions we ask. Analogically, if the hammer is chosen for setting screws, timbers will split. The tool must comport with the question, or else the tool becomes as much a hindrance as anything. It turns out that things in the human world are rarely linear, and things in the *criminological* world are rarely as simple as direct, even nonlinear, quantitative relations require. That said, there remains an important place for statistical analyses like those discussed here in the human sciences, among other ways of knowing. Practical programmatic questions related to performance and outcomes, for example, lend themselves to quantification, although fuller understandings tend to require analyses in addition to quantitative ones. Further, quantitative human sciences that ask "whether-or-not" questions (e.g., Ziliak & McCloskey, 2011) are still being posed in the human sciences and not without merit, although in theoretical contexts, the objects of study are abstractions of abstractions inferred to exist on the strength of third-order indirect evidence, and we forget that to our peril. Statistical reasoning from CLT in the human sciences is one of the many useful tools in the now-growing human sciences toolbox ... when used according to its capacity.

Well, that was a lot of wind to get us to this point. Most of the techniques we have discussed thus far operate from a leveraging of inference powered by the Central Limit Theorem, allowing us to draw conclusions about a population on the basis of analyses of an unbiased sample. For them to work as intended, certain assumptions are made about the nature of the population, and about the samples drawn from them. You probably don't have to be convinced that CLT-driven inference is not the whole ball game when it comes to learning things about the human condition. Oftentimes, it is as useful merely to know something about the sample before us, and to leave inference about the applicability of findings to other settings in time and place to one's audience of other researchers.

It shouldn't surprise, then, that there is a whole bank of techniques that do not lean on CLT-powered inference, tests that don't assume certain things to be true about the distribution of variables, tests that represent rather clever, even pedestrian reasoning systems to answer questions. Oh, sure, they may make connections between samples and populations, and some will put probability curves to use in their calculations, but the sample–population connection is not one based on assumptions about the distributions of the variables (and thus their representativeness of the sampling distribution of the population).

Oftentimes, **nonparametric** tests provide analytical techniques that are not as hindered by "burdensome" assumptions, and as a result, they may well enjoy both a longer productive lifespan and also a wider compass of usefulness than more sophisticated multivariate techniques. Nonparametric techniques offer answers to questions that researchers tend to pose, but these techniques allow for some more latitude regarding the condition of the data fed them. In all fairness, t test and ANOVA are robust techniques, and it is highly unlikely a careless researcher will break the tool by feeding it marginal data (e.g., summed Likert-scale indices being treated as truly metric variables, for example, or variables with moderate skew), but nonparametric techniques do not make the same demands on data. At the same time, nonparametric techniques can give a measure of insight into data patterns. They are not as fancy, and they surely are not as efficient as other strategies, but again, analytical sophistication should not necessarily be the first priority when doing sound human science. As a result of their humbleness, nonparametric techniques tend to live on the margins of town and mostly keep to themselves. Researchers and practitioners would do well, though, to visit that side of town more often than they do currently. All the tests discussed in this chapter are, of course, nonparametric.

The location of this chapter at the end of the treatment on techniques should not be taken as a de-prioritization of nonparametric analytical techniques. Most statistical texts locate a chapter or two on nonparametric techniques at the back of the book. To be forthright with you, in writing this text, I grappled with this choice myself; having taught statistics many times, I feel that some of these techniques are wonderfully useful pedagogically as well as analytically. They present a kind of reasoning that is found in statistics elsewhere, and as I've said, they are not overlain with assumptions about hypothetical sampling distributions. I chose to put the chapter here at the end because the rest of what has come before operates, I hope, as something of a narrative that builds on itself from standard deviation through to the Central Limit Theorem, up to the chapter on regression (and beyond into other works on statistics). As this chapter represents something of a break in that narrative, it seems appropriate to place it here at the end. We might even choose to assume, though, that the best has been saved for last.

Of course, anyone who chooses to use this resource for their own teaching is encouraged to introduce this chapter at whatever point that suits them, and in fact, I have taught using this book with this chapter falling earlier in the course narrative, to great effect. I leave it to you, gentle reader, to decide how you most prefer to consume and utilize this work.

Chi Square (χ^2)

Perhaps no observation I could offer can make the case for the capacity of nonparametric statistical techniques quite so well as the example of chi square (χ^2). As an analytical technique, χ^2 certainly has its limitations, but it operates powerfully in the space in which it works appropriately.

Simply, χ^2 is a test of the goodness of fit between frequencies in a contingency table relative to expected values. First, let's talk about a contingency table, and then we will get to the rest.

A contingency table is a rudimentary grouped frequency analysis in tabular form. It is an analysis because it does not present raw scores, but instead portrays counts of cases (or events or things) according to categories. It is grouped frequencies because, well, it presents frequencies ... in groupings. Table 13.1 is a fine example of a contingency table. When we talk about such tables, it is common parlance to describe it according to the number of rows and the number of tables (in that order – *row* × *column*). Table 13.1, then, is a 2 × 3 contingency table.

Typically, we tend to place the dependent variable in columns and the independent variable in rows. That makes sense here, as the security level is unlikely to predict sex (although if we need to, we can frame the question that way). In this case, we are going to hypothesize that there is a relationship between sex and the assigned security level: males are more likely

Table 13.1 Security Level for Hypothetical Intake Unit, February 2019

		Security Level		
		Low	*Medium*	*High*
Self-reported sex	Male	35	47	19
	Female	38	12	10

Table 13.2 Security Level for Hypothetical Intake Unit, February 2019 with Totals

		Security Level			*Totals*
		Low	*Medium*	*High*	
Self-reported sex	Male	35	47	19	101
	Female	38	12	10	60
	Totals	73	59	29	161

to be assigned a higher security level than females. The null hypothesis, then, is that there is no relationship between sex and security level.

In Table 13.1, we can see both rows and columns, with each cell representing a portion of the total cases – in this case, the total number of prison intakes for a hypothetical intake unit in a hypothetical state. As it turns out, there were 161 internees in-processed that month (adding up all the numbers in the cells gave us this number), of which 101 were males and 60 were females (we added up all the numbers across in one row for each group). We can also say that of the 161 internees in-processed, 73 (35 + 38) were into low-tier security, 59 (47 + 12) were assigned to a medium-level security status, and 29 (19 + 10) were placed in high security. Let's add this other stuff in to the table:

In Table 13.2, we can see not only row and column totals (what we call row and column *marginal* totals, or row and column marginals) set in light gray, we also have the grand total number of cases, set in darker gray. If you add either of the sets of row marginals or column marginals, you should get the same grand total (73 + 59 + 29 = 161, and 101 + 60 = 161, so we nailed it).

The Calculation of χ^2

It turns out that the calculation of χ^2 is exceedingly basic, although it requires a few steps. The formula for χ^2 is:

$$\chi^2 = \sum \frac{(o-e)^2}{e}$$

where "*o*" stands for our observed scores (meaning those frequency scores in the boxes in Table 13.1) and "*e*" stands for our expected scores for each of these cells. Essentially, then, χ^2 is the difference between observed and expected scores, squared, divided by expected scores. There's just no magic here (which means all the magic is in you).

But hold up … If we knew what values were expected, wouldn't we be in a place where we didn't really need all of this? Fair point. More, if we *can* approximate expected values, how would we go about doing such a thing? Also a fine question.

Expected frequency values can be introduced or imputed from a variety of places; we might choose expected values from previous studies, from theory, from other literature, from program thresholds, and we can also calculate expected values in a number of ways. The way we will do it here is through something of an averaging process. Expected frequencies for each cell can be derived:

$$f_{ex} = (r_m \times c_m)/\text{total}$$

Or in English, an expected frequency value for any given cell is calculable by taking the cell's row marginal and multiplying it by its column marginal, and then dividing that product by the total. Each cell contributes to only one column marginal total and one row marginal total. These marginal totals are multiplied, and the product is divided by the total count.

The null hypothesis is that the summed difference between observed and expected values is 0. The indication would be that there is a perfect fit between observed and expected scores, and thus there is no meaningful impact from independent variable (usually the variable presented in rows) group membership. In our example, the null hypothesis is that we cannot predict self-reported sex knowing only the security classification to which an internee was assigned at intake. Table 13.3 illustrates how we will calculate each cell's expected value.

If we do the arithmetic indicated in Table 13.3, we get the values shown in Table 13.4.

Now we have both a complete set of observed scores (those are the ones we were given in this case) and a complete set of expected scores (those are the scores in Table 13.4). Next, we have to find the difference, subtracting expected from observed, and then squaring the difference.

Table 13.3 Calculation of Expected Values

		Security Level			*Totals*
		Low	*Medium*	*High*	
Self-reported sex	Male	(73 × 101)/161	(59 × 101)/161	(29 × 101)/161	101
	Female	(73 × 60)/161	(59 × 60)/161	(29/60)/161	60
	Totals	73	59	29	161

Table 13.4 Expected Values, Completed

		Security Level			*Totals*
		Low	*Medium*	*High*	
Self-reported sex	Male	45.795	37.012	18.193	101
	Female	27.205	21.988	10.807	60
	Totals	73	59	29	161

Right away you might be detecting something of a theme. Yeah, the differences end up patterning in some fashion, because in essence, cases are moving from one place as expected scores to another place as observed scores. The differences have to go somewhere. If you were to add the differences across a row or down a column, they will sum to 0. Slick, right? Now we continue by squaring the differences we just found:

We are almost there. One more major step, and that is to divide the squared difference from Table 13.6 by the expected values from Table 13.4. Table 13.7 presents the results.

Finally, we sum all cells and get our final χ^2_{obs} value: 2.545 + 4.283 + 2.695 + 4.537 +.036 + .060 = 14.156. Of course, we will need a χ^2_{crit} value to which to compare it. The chi-square distribution is in essence one-half of a normal distribution, so values will always be positive.

For the determination of χ^2_{crit}, though, we need to calculate degrees of freedom, and then reference the chi-square table (Appendix D). The degrees of freedom for chi square is calculated as follows:

$$\text{df} = (\text{rows} - 1)(\text{columns} - 1)$$

In this instance, 2 – 1 rows, times 3 – 1 = 2 columns, 2 × 1 = 2 degrees of freedom. For p = 0.05, our χ_{crit}^2 is 5.99. Because our χ_{obs}^2 (14.156) exceeds χ_{crit}^2 (5.99), we can reject the null hypothesis.

Table 13.5 Observed Minus Expected

		Security Level		
		Low	*Medium*	*High*
Self-reported sex	Male	35 – 45.795 = –10.795	47 – 37.012 = 9.988	19 – 18.193 = 0.807
	Female	38 – 27.205 = 10.795	12 – 21.988 = –9.988	10 – 10.807 = –0.807

Table 13.6 Squaring the Difference

		Security Level		
		Low	*Medium*	*High*
Self-reported sex	Male	-10.795^2 = 116.532	9.988^2 = 99.760	0.807^2 = 0.651
	Female	10.795^2 = 116.532	-9.988^2 = 99.760	-0.807^2 = 0.651

Table 13.7 Numerator Divided by Denominator

		Security Level		
		Low	*Medium*	*High*
Self-reported sex	Male	116.532/45.795 = 2.545	99.760/37.012 = 2.695	0.651/18.193 = 0.036
	Female	116.532/27.205 = 4.283	99.760/21.988 = 4.537	0.651/10.907 = 0.060

Table 13.8 Patterning in the Differences

		Security Level		
		Low	*Medium*	*High*
Self-reported sex	Male	35 − 45.795 = −10.795 Observed less than expected	47 − 37.012 = 9.988 Observed more than expected	19 − 18.193 = 0.807 Observed and expected about the same
	Female	38 − 27.205 = 10.795 Observed more than expected	12 − 21.988 = −9.988 Observed less than expected	10 − 10.807 = −0.807 Observed and expected about the same

Interpretation

Looking back at Table 13.5 in which we compared observed and expected scores, we can see that there is something of a pattern – a pattern made explicit in Table 13.8. The diagonal cells for males in low-level security and the females in medium-level classification were lower than expected according to our averaged expected cell frequencies; fewer men than were expected ended up in the low group and fewer women than expected ended up in the middle group. Consequently, more females than expected were in the low classification and more men than expected were in the medium classification. These differences, as we have seen, are statistically nontrivial.

If we had a circumstance where we had a 2 × 2 contingency table, and if we sought more than a goodness-of-fit outcome, we could perform phi, as we have discussed.

It bears observing at this point that χ^2, useful though it may be, suffers from rather important limits, most of which stem from the mechanics of adding differences across a number of cells. If there are lots of cells, or if the total sample size is greater than 120, relatively minor differences can add up to be sufficiently large so as to be statistically significant without being operationally meaningful. Another way to say this is that in large-*n* environments, χ^2 can be overly sensitive to small differences. In cases where cell frequencies get to be *small*, say less than 5, χ^2 gets a little wonky as well. Low expected values in cells, particularly in circumstances where there are a relatively large number of cells and also high expected values in other cells means that χ^2 is losing whatever signal is present in the louder noise of the empty or near-empty cells.

For high-end problems, one could (a) find another test (restructure the question or the data); (b) randomly select subsamples and perform χ^2 on them. For low-end problems, one can use Fisher's exact test, but it works better with relatively few cells (and runs automatically in some statistical software for 2 × 2 tables).

Measures of Association to Accompany χ^2: phi and Cramer's V

Recall from Chapter 11 that phi is a measure of association. It is useful with χ^2 to determine the strength of the relationship between two variables. Remember, χ^2 only speaks to the goodness of fit between the two variables. It does not give us an inkling into the strength of the relationship. For such a sense of things, we need other measures, specifically phi and Cramer's *V*. Phi is exceedingly simple to calculate:

$$\phi = \sqrt{\frac{\chi^2}{n}}$$

but it is suited only for situations involving 2 × 2 tables. For any other situation, we need to rely on the similar measure, Cramer's *V* (Bachman & Paternoster, 2017: 260):

$$V = \sqrt{\frac{\chi^2}{n(k-1)}}$$

Where *k* is the number of rows *or* the number of columns, whichever is *smaller* (e.g., Bachman & Paternoster, 2017). Now, because our example produced an observed χ^2 that was statistically significant, we should take the next step and find the appropriate measure of relationship strength for our example. Table 13.9 is just Table 13.4 again, provided here for your convenience. Our χ^2_{obs} was 14.156, if you recall.

OK, so first, because the matrix is a 2 × 3 matrix, phi is ruled out, and Cramer's *V* is the choice (if you always used Cramer, though, it wouldn't matter, as it would be the same for 2 × 2 situations anyway, as you'll see). For *k*, though, we have to find the lesser number of rows or of columns. In this case, as there are only two rows and there are three columns, the number of rows is our pick. So, *k*, then, is 2.

$$V = \sqrt{\frac{14.156}{161(2-1)}}$$

As you might already see, the numerator will be the same as it would have been for phi.

$$V = \sqrt{\frac{14.156}{161(1)}}$$

$$V = \sqrt{\frac{14.156}{161}}$$

$$V = \sqrt{\frac{14.156}{161}}$$

Table 13.9 Expected Values, Completed

		Security Level			*Totals*
		Low	*Medium*	*High*	
Self-reported Sex	Male	45.795	37.012	18.193	101
	Female	27.205	21.988	10.807	60
	Totals	73	59	29	161

$$V = \sqrt{.08793}$$

$$V = 0.2965$$

which, in terms of scale, is perhaps on the low end of a moderate-strength relationship.

Nonparametric Proportionate Reduction in Error (PRE) Measures

In simplest terms, if we can derive some form of improved estimate of the value of the dependent variable by using a value of the independent variable relative to any guess for the value of the dependent value based on the mean of that dependent variable, we've achieved a reduction in error. We can ask, expressing the same idea, "to what degree does our knowledge of *X* help us increase our knowledge of *Y*?" Actually, you may recognize that little bit of reasoning from bivariate regression; to what degree does *X* help us deflect away from the mean of *Y* as the most capable predictor of the value of any *Y*? We did this sort of reasoning using a regression line in Chapter 12. In a rough sense, r^2 is akin to a kind of PRE measure, although not in every respect (Costner, 1965; Kviz, 1981). Costner (p. 344) set up four criteria for PRE measures:

1 There must be a rule for estimating values of the dependent variable utilizing values of the independent variable ("The Informed Decision Rule" [see Weisburd & Britt, 2008]);
2 There must be a rule for estimating the same values of the dependent variable as above without utilizing values of the independent variable ("The Naïve Decision Rule");
3 Some notion of "error" and its measurement; and
4 A measure of association that takes the form:

$$\text{PRE} = \frac{\text{Error by naive decision rule} - \text{error by informed decision rule}}{\text{error by naive decision rule}}$$

OK, so what does all of that mean? It means that we are seeking to be able to understand something about the variability in our dependent variable based on the variability in the independent variable.

Conceptually speaking, the numerator represents the difference in precision made by the informed decision rule, and the denominator is the baseline, the amount of error using, say, the mode of the dependent variable. That makes the PRE standard measure of association a ratio; a value of 1 means that the informed decision rule explains it all and is fully revealing regarding variability of the dependent variable. A value of 0 means that the informed decision rule doesn't do any better than the measure of centrality of the dependent variable; it means, then, that there's no *there* there.

Now, it clearly matters, then, which variable is identified as the dependent variable. This determination can come from theory or from previous research (e.g., Weisburd & Britt, 2008), but it should be made in advance of analyses.

As mentioned earlier, these PRE measures are for use with nominal or ordinal variables (see, e.g., Crittenden & Montgomery, 1980), and as it turns out, there are a few of each. Because of the scope and purpose of this work, we will only treat in detail two – one each for nominal and for ordinal variables.

Table 13.10 Crosstabulation of Information Receipt and Age for a Hypothetical Sample of Community Watch Participants

	Routinely Misses PD Information	*Routinely Receives PD Information*	*Marginal Totals*
Computer literate	38	88	126
Not computer literate	75	54	129
Marginal totals	113	142	255

PRE Measures for Nominal Variables

Goodman and Kruskal's gamma and lambda are the standard PRE measure for nominal variables, and so I am happy to arrange an introduction here. Let's begin with our setting.

The police department hosted a cookout and community softball game in a neighborhood, and attendance wasn't all that the chief had hoped. After attending several community watch meetings, you began to sense something: some people in the community served by your police department do not seem to be getting announcements for events, or to be benefiting from PSAs provided through the police department's social media services. Community member utilization of the police department's social media services might vary as a function of citizens' computer literacy. You arranged a brief paper-and-pencil survey for city residents, administered at community watch meetings: the survey contained a few questions about police department events and also about computer literacy. Your results looked like Table 13.10.

We are going to determine the degree to which computer literacy helps us adequately predict who routinely is missing out on police department information. Again, we need to be clear about which variable is our dependent variable, and which is serving as the independent variable. In this case, the information variable is our dependent variable, whereas the computer literacy measure is our independent variable; we think computer literacy can predict in part the getting (or not getting) of department information.

Goodman and Kruskal's Lambda

First, we will use lambda to treat the question of whether knowing something about computer literacy (our independent variable) can reveal something about how informed a person is regarding police department information. And again, like with χ^2, lambda will be operated from a contingency table. Knowledge of the independent variable may well help us successfully "predict" outcomes, that is, different cell frequencies, for levels of the dependent variable.

Computationally, lambda (λ) looks like this (Bachman & Paternoster, 2017: 261):

$$\lambda = \frac{\left(\sum f_i\right) - f_d}{n - f_d}$$

where f_i is the biggest *cell frequency* for each level of the independent variable (that is, the modal cell value for each row), f_d is the biggest *marginal total* for the dependent variable (that is, the biggest column marginal total) and n is the grand total.

As per usual, our dependent variable (regularly catching or missing PD information) is in columns, and our independent variable (computer literacy) is in rows. The biggest cell frequency for the computer-literate level is 88, and the biggest cell frequency for the non-literate level is 75. The biggest marginal total for the dependent variable is 142 – the group that routinely received PD information. If we were merely to assume that all 255 people routinely received informational updates from the police department, we would be right 142 times and wrong 113 times.

$$\lambda = \frac{\left(\sum f_i\right) - f_d}{n - f_d}$$

$$\lambda = \frac{(88+75)-142}{255-142}$$

$$\lambda = \frac{163-142}{113}$$

$$\lambda = \frac{21}{113}$$

$$\lambda = 0.1858$$

which tells us that knowledge of the independent variable (literacy) can reduce our predictive errors for the dependent variable by about 18.6% (0.1858 × 100 will give us a percentage PRE).

We can also think about lambda as being a *verbatim* version of the fourth element in the definition of PRE:

$$\text{PRE} = \frac{\text{naive error} - \text{informed error}}{\text{naive error}}$$

or:

$$\lambda = \frac{E_1 - E_2}{E_1}$$

where E_1 refers to the errors made in prediction when we ignore information from the independent variable (that is, the grand non-modal frequency), and E_2 refers to the errors made when we apply information from the dependent variable (or the non-modal frequencies for each level of the independent variable).

Said another way, we would write:

$$\lambda_B = \frac{P_e - P_{e|A}}{P_e}$$

where P_e is the probability of error, and $P_{e|A}$ is the probability of error given information from variable *A*.

Now notably, the obverse isn't the same – rather, the ability of variable *A* to reduce errors of "prediction" for variable *B* is not the same as the ability of variable *B* to reduce prediction errors for variable *A*. In other words, lambda is asymmetric. Gamma, though, isn't, and that's what we'll discuss next.

Goodman and Kruskal's Gamma

Gamma (γ) works for nominal or ordinal variables, but we will discuss it here. Of course, it follows the same basic logic we've been using with lambda, but it takes a slightly different shape:

$$\gamma = \frac{P_c - P_d}{P_c + P_d}$$

such that P_c stands for concordant pairs, and P_d represents discordant pairs. Again, with γ, we are working in a contingency table like others we've been using in this chapter. Actually, let's reflect back on χ^2 for a second. Recall in Table 13.8 how we saw a kind of patterning happening? Depending on the relationship between the variables, one might find a kind of shape to which cells had higher frequencies than expected. In a 2 × 2 table, it might look like this (Table 13.11):

We can utilize this same patterning to determine the degree to which a relationship exists between the two variables. As indicated, we will think about this patterning in terms of concordant and discordant pairs. Now, with a 2 × 2 table like this one, the determination is simple (Table 13.12):

A and *C* are concordant whereas *B* and *D* are discordant. If no relationship existed between the two variables (here, that is, the variables would be sex and some binary, yes/no question), we might find a contingency table that looked like this:

In this instance, a perfect relationship might look like this:

Numbers dance around the contingency table according to some hidden gravity, and it is this gravity that we seek to understand, although again, our analyses can only ever reveal the possibility of the existence of the forces of interest, and not their truer nature.

For a two-by-two table like this one, we have a relatively simple process for gamma:

$$\gamma = \frac{P_c - P_d}{P_c + P_d}$$

$$\gamma = \frac{AC - BD}{AC + BD}$$

Table 13.11 Hypothetical Contingency Table

	Yes	*No*
Male	More than expected	Less than expected
Female	Less than expected	More than expected

Table 13.12 Hypothetical Contingency Table

	Yes	*No*
Male	*A*	*B*
Female	*D*	*C*

Table 13.13 Hypothetical Contingency Table

	Yes	*No*
Male	25	25
Female	25	25

Table 13.14 Hypothetical Contingency Table

	Yes	*No*
Male	50	0
Female	0	50

For Table 13.13, the result would look like this:

$$\gamma = \frac{AC - BD}{AC + BD}$$

$$\gamma = \frac{(25 \times 25) - (25 \times 25)}{(25 \times 25) + (25 \times 25)}$$

$$\gamma = \frac{625 - 625}{625 + 625}$$

$$\gamma = \frac{0}{1{,}250}$$

$$\gamma = 0$$

In contrast, Table 13.14 would give us:

$$\gamma = \frac{AC - BD}{AC + BD}$$

$$\gamma = \frac{(50 \times 50) - (0 \times 0)}{(50 \times 50) + (0 \times 0)}$$

$$\gamma = \frac{2{,}500 - 0}{2{,}500 + 0}$$

$$\gamma = \frac{2{,}500}{2{,}500}$$

$$\gamma = 1$$

Yep, you guessed it: like other measures of association, gamma yields a result between –1 and +1, with a 0 outcome indicating no relationship, a +1.0 indicating a perfect direct relationship, and –1 indicating a perfect inverse relationship. Flip the 50s and the 0s in Table 13.14 and you'll see what I mean.

For a 2 × 2 table like these, gamma is actually also something called Yule's Q, but we won't spend more time there so that we can move on to more involved calculations of gamma. Let's return to the example we used when we were talking about χ^2 (Table 13.15).

Table 13.15 Security Level for Hypothetical Intake Unit, February 2019 with Totals

		Security Level			*Totals*
		Low	*Medium*	*High*	
Self-reported sex	Male	35	47	19	101
	Female	38	12	10	60
	Totals	73	59	29	161

Here, the *AC* – *BD* business can't work because there are more than four cells, so we have to add a few steps. The process is simple, but it requires a little care and attention to detail. First, we will find the total value for concordant pairs, and then use the same process in reverse to find the total for discordant pairs.

The Snacky Statistician: Cooking a Goodman and Kruskal's Gamma

Step 1. find the top-leftmost cell. Here, that cell (Male/Low Security) has a value of 35.
Step 2. find *all* of the cells that are below and to the right of that cell. In this case, that would be Female, both Medium and High Security, and add them up. In this case, we would add 12 and 10.
Step 3. multiply 35 times the sum of the cells below and to the right; again, 12 and 10. That gives us 35(12 + 10), which is 35(22) or 770.
Step 4. slide one cell to the right of the first one you used – in this case that would be Male/Medium Security = 47.
Step 5. just as before, look to the right and below, and add up all cells. In this instance, we would only have Female/High Security (=10).
Step 6. Do the same math as before: 47(10) = 470.
Step 7. The last cell reading rightward in the top row does not have a cell below and to its right, so you're done with that row.
Step 8. Go to the next row down from the first cell. Here that's Female/Low Security. Are there any cells below and to its right? Nope. You're done with concordant pairs. Add up all of the values: 770 + 470 = 1240.
Step 9. beginning with the top right cell, we will work the same process, only backward (that is, we will add all the cells below and to the left of it). Thus, we will have 19(12 + 38) = 950.
Step 10. slide over to the left to Female/Medium Security, and then sum all the cells below and to its left. We will get 47(38) = 1,786.
Step 11. Add all of the discordant values together: 950 + 1786 = 2,736.
Step 12. find Gamma.

Serve immediately.

$$\gamma = \frac{P_c - P_d}{P_c + P_d}$$

$$\gamma = \frac{1,240 - 2,736}{1,240 + 2,736}$$

$$\gamma = \frac{-1,496}{3,976}$$

$$\gamma = -0.3762$$

In this case, we have found that there is a moderate inverse relationship between sex and security classification. More specifically, recall that females were more likely than expected (via χ^2) to receive a medium security classification, whereas males were somewhat less likely than expected to end up with a low-security classification. This finding comports with our χ^2 results: 14.156 was our observed χ^2, which enabled us to reject the null hypothesis, and Cramer's *V* (0.2965) also indicated the existence of a moderate-strength relationship (although *V* does not indicate directionality the way gamma does).

Let's do one more example for both lambda and gamma using a larger contingency table. Lambda is intended for use with nominal variables, or for variables that can be treated as if they're nominal (which isn't quite the same thing).

A poll was taken regarding whether the police officer union should seek a more enhanced dental insurance package for employees. The hypothetical results, separated by shift assignment, are shown in Table 13.16.

Your thought was that shift assignment might impact whether employees supported paying more for more comprehensive dental coverage. In this instance, shift assignment is the independent variable, and opinion regarding dental coverage is the dependent variable.

Let's do lambda first. The independent variable – shift assignment – is nominal. Easy enough. Although the dependent variable is best understood as ordinal, it can also be treated as if it is nominal (opinions are semi-discrete things; levels of agreement are not necessarily related in an orderable fashion).

$$\lambda = \frac{\left(\sum f_i\right) - f_d}{n - f_d}$$

Table 13.16 Agreement Regarding the Need for Enhanced Dental Insurance by Shift

	Strongly Disagree	*Disagree*	*Agree*	*Strongly Agree*	*Totals*
Day shift	14	18	10	9	51
Evening shift	10	11	13	16	50
Night shift	8	10	15	20	53
Totals	32	39	38	45	153

Or modal totals for each row are 18, 16, and 20. In these, we have identified the largest cell value for each level of the independent variable. Summed, they equal 54.

The modal category for the dependent variable is "Strongly Agree" and the corresponding value for that column is 45. Our n, of course, is 153. We can roll.

$$\lambda = \frac{(18+16+20)-45}{153-45}$$

$$\lambda = \frac{(54)-45}{153-45}$$

$$\lambda = \frac{9}{108}$$

$$\lambda = 0.0833$$

Not a fantastically large benefit: knowing someone's shift assignment decreases our error regarding their feelings about dental coverage by a little more than 8%. Now let's do gamma.

$$\gamma = \frac{P_c - P_d}{P_c + P_d}$$

To derive concordant and discordant pairs will require several steps. I've tried to make the steps a little clearer by highlighting the relevant cells in Tables 13.17–13.22.

Table 13.17 Agreement Regarding the Need for Enhanced Dental Insurance by Shift

	Strongly Disagree	*Disagree*	*Agree*	*Strongly Agree*	*Totals*
Day shift	14	18	10	9	51
Evening shift	10	11	13	16	50
Night shift	8	10	15	20	53
Totals	32	39	38	45	153

Table 13.18 Agreement Regarding the Need for Enhanced Dental Insurance by Shift

	Strongly Disagree	*Disagree*	*Agree*	*Strongly Agree*	*Totals*
Day shift	14	18	10	9	51
Evening shift	10	11	13	16	50
Night shift	8	10	15	20	53
Totals	32	39	38	45	153

Table 13.19 Agreement Regarding the Need for Enhanced Dental Insurance by Shift

	Strongly Disagree	*Disagree*	*Agree*	*Strongly Agree*	*Totals*
Day shift	14	18	10	9	51
Evening shift	10	11	13	16	50
Night shift	8	10	15	20	53
Totals	32	39	38	45	153

Table 13.20 Agreement Regarding the Need for Enhanced Dental Insurance by Shift

	Strongly Disagree	*Disagree*	*Agree*	*Strongly Agree*	*Totals*
Day shift	14	18	10	9	51
Evening shift	10	11	13	16	50
Night shift	8	10	15	20	53
Totals	32	39	38	45	153

Table 13.21 Agreement Regarding the Need for Enhanced Dental Insurance by Shift

	Strongly Disagree	*Disagree*	*Agree*	*Strongly Agree*	*Totals*
Day shift	14	18	10	9	51
Evening shift	10	11	13	16	50
Night shift	8	10	15	20	53
Totals	32	39	38	45	153

Table 13.22 Agreement Regarding the Need for Enhanced Dental Insurance by Shift

	Strongly Disagree	*Disagree*	*Agree*	*Strongly Agree*	*Totals*
Day shift	14	18	10	9	51
Evening shift	10	11	13	16	50
Night shift	8	10	15	20	53
Totals	32	39	38	45	153

Recall how we had to begin with the top left cell and then find all of the values below and to the right? Well, in this case, we need to grab six values. That means our first set of concordant pairs is:

$$14(11+13+16+10+15+20)=1{,}050$$

Next, we would find the concordant pair value for the next cell over: 18. Thus, we would have

$$18(13+16+15+20)=1{,}152$$

Our final bit of information regarding concordant pairs is as can be seen in Table 13.19:

$$10(16+20)=360$$

And now for the discordant pairs. Working in the opposite direction, we would find:

$$9(10+11+13+8+10+15)=603$$

And then...

$$10(10+11+8+10)=390$$

And finally...

$$18(10+8)=324$$

Concordant pairs:

$$14(11+13+16+10+15+20)=1{,}050$$

$$18(13+16+15+20)=1{,}152$$

$$10(16+20)=360$$

$$1{,}050+1{,}152+360=2{,}562$$

Discordant pairs:

$$9(10+11+13+8+10+15)=603$$

$$10(10+11+8+10)=390$$

$$18(10+8)=324$$

$$603+390+324=777$$

$$\gamma=\frac{P_c-P_d}{P_c+P_d}$$

$$\gamma=\frac{2{,}562-777}{2{,}562+777}$$

$$\gamma=\frac{1{,}785}{3{,}339}$$

$$\gamma=0.5346$$

Interestingly, gamma seems to indicate a moderately strong relationship whereas lambda does not. Remember that lambda is asymmetrical, whereas gamma isn't. Gamma tells us that shift assignment is related to support for better dental coverage; moving from day to evening to night shifts, it seems agreement increases somewhat (concordance – moving left-to-right and downward – is greater than discordance – moving right-to-left and downward). Lambda, though, seems to indicate that the amount of help lent by shift assignment is relatively small: knowing shift assignment helps us predict support for enhanced dental care by just a bit over 8% than without that knowledge.

One Last Tool: Median Sign Test

This bears repeating: not all statistical techniques involve fancy analytical strategies. What they do tend to represent, though, is a clever problem-solving strategy appropriate for particular circumstances. Some, like this median sign test, feel almost like the kind of thing that an attentive lieutenant cooked up on the fly. The median sign test is an intuitive and simple strategy for speaking in precise and contextualized terms about a circumstance regarding the difference between actual and expected outcomes. It is just this innovative spirit that I have hoped to cultivate throughout this text.

The median sign test (e.g., Kachigan, 1986) utilizes two conceptual tools with which by now we have become rather familiar: a comparison through subtraction between observed and expected scores (like in *t* test and χ^2), and a contextualization of outcome using a probability curve. The differences here, though, are what information we put to use from the comparisons, and the nature of the contextualization process.

Say the chief has directed foot-patrol officers to engage in and document conversations with citizens. The chief does not want to know with whom officers are speaking or about the topics they discuss, but he is anxious to make sure that officers are engaging in face-to-face, friendly engagements outside of the law-enforcement framework. Let's say officers are asked to report to sergeants the number of conversations they strike up during a regular day or evening shift. Let's also say that the chief wants each officer to have 20 brief conversations in a regular foot-patrol day or evening shift. Below are the counts of conversations from each of the $n = 10$ officers on last night's evening shift:

22, 16, 30, 21, 14, 9, 11, 15, 12, 14

A very simple way to treat the differences between the above observed scores and the expected value of 20 conversations is to do the subtraction (observed minus expected), and then record just the positive or negative sign for the result, like this:

+, –, +, +, –, –, –, –, –, –

Then we would count the plus signs (here we had three plus signs out of ten). If we had an officer who had exactly 20 conversations, the difference would be 0, so no sign would result. In that case, we would delete it from the analysis and operate with the scores that remain.

In order to derive a measure of context for interpretation purposes, we can use the binomial theorem:

$$P(x) = \frac{n!}{r!(n-r)!} p^r q^{n-r}$$

to determine the likelihood of *x* successes on y attempts. As it happens, we already calculated the probabilities for a ten-trial binomial outcome:

Coin flips or value signs, it is all the same to you and me, at least in terms of defining success and trials. Looking at Table 13.23 (which is otherwise identical to Table 5.2), we can see that the probability of getting three successes on ten trial rounds to 0.1172. Thus, by chance alone (assuming, that is, that each officer's compliance was as likely as achieving a heads on a single coin flip, which is a rather charitable assumption), we could expect to have only three officers out of ten exceed the number of contacts desired by the chief around 11.7% of the time. If all things are equal, then, 12 shifts out of 100 day or evening shifts should have only three out of ten officers exceed the threshold sought. Whether that is statistically trivial or not for you is a thing to determine in advance (in the body of a critical value standard), but at the least, it might be worth a conversation with that sergeant before

Table 13.23 Probability Mass Function for Discrete Bernoulli Outcomes for a Fair Coin, Ten Flips

Outcome	*Count*	*Probability*	*P(x)*
0	1	0.000977	0.000977
1	10	0.000977	0.009766
2	45	0.000977	0.043945
3	120	0.000977	0.117188
4	210	0.000977	0.205078
5	252	0.000977	0.246094
6	210	0.000977	0.205078
7	120	0.000977	0.117188
8	45	0.000977	0.043945
9	10	0.000977	0.009766
10	1	0.000977	0.000977

the chief gets too angry. That, or perhaps officers will simply start reporting conversations that never happened, now that they know the "target" number sought by the chief.

Summary

All that has come before constitutes something of a gradual crescendo into the chapter on bivariate regression, Chapter 12. Well, as important as the Central Limit Theorem and all the rest is for the kind of reasoning we've been seeking to be able to do, it turns out that if we set aside all of that reasoning for a moment, we can make use of another set of techniques. Nonparametric statistical techniques have considerable capacity without some of the impedimenta of demanding assumptions. Notably, though, these techniques come with their own set of capacities and limitations, but the utility of chi square, Cramer's *V*, lambda, gamma and the median sign test may well outstrip fancier multivariate techniques in the fullness of time, at least in the context of human sciences.

Main Take-Aways

- Chi square, *V*, lambda and gamma operate from reduced data – specifically, they are used on contingency tables.
- Chi-square tests the goodness of fit between observed and expected scores.
- Phi and Cramer's *V* test the strength of a statistical relationship.
- PRE techniques, like lambda and gamma, work to determine the degree to which information contained in one variable helps us reduce "error" in another variable.
- All of these techniques represent rather clever approaches to answering intuitive if basic questions about relationships within contingency tables.

Key Terms (in order of appearance)

nonparametric statistics	chi square	Cramer's *V*
proportionate reduction in error	Lambda	Gamma
Median sign test		

References

Adam, E., Quinn, M., Tavernier, R., McQuillan, M., Dahlke, K., & Gilbert, K. (2017). Diurnal cortisol slopes and mental and physical health outcomes: A systematic review and meta-analysis. *Psychoneuroendocrinology 83*, 25–41. https://doi.org/10.1016/j.psyneuen.2017.05.018.

Argyris, C. & Schon, D. (1974). *Theory in practice: Increasing professional effectiveness.* Hoboken, NJ: Jossey-Bass.

Ashby, M. (2024). Why you can't identify changes in crime by comparing this month to last month. *Social Research Association.* Accessed Monday, February 19, 2024 from https://the-sra.org.uk/SRA/Blog/whyyoucantidentifychangesincrimebycomparingthismonthtolastmonth.aspx

Bachman, R., & Paternoster, R. (2017). *Statistics for criminology and criminal justice, 4d.* Thousand Oaks, CA: Sage.

Bedeian, A., & Mossholder, K. (2000). On the use of the coefficient of variation as a measure of diversity. *Organizational Research Methods 3*, 3, 285–298.

Buscariolli, A. (2023). Making crime visible in the digital age: The ethnomethods of data policing. *Critical Criminology.* https://doi.org/10.1007/s10612-023-09683-y.

Cay, M., Ucar, C,m Senol, D., Cevirgen, F., Ozbag, D., Altay, Z., & Yildiz, S. (2017). Effect of increase in cortisol level due to stress in healthy young individuals on dynamic and static balance scores. *Northern Clinics of Istanbul 5*, 4, 295–301.

Chan, S., & Debono, M. (2010). Replication of cortisol circadian rhythm: New advances in hydrocortisone replacement therapy. *Therapeutic Advances in Endocrinology and Metabolism 1*, 3, 129–138.

City of Detroit's Open Data Portal (2020) https://data.detroitmi.gov/datasets/detroitmi::rms-crime-incidents/about

Cohen, J. (1988). *Statistical power analysis for the behavioral sciences.* Hillsdale, NJ: Lawrence Erlbaum.

Costner, H. (1965). Criteria for measures of association. *American Sociological Review 30*, 3, 341–353.

Crittenden, K., & Montgomery, A. (1980). A system of paired asymmetric measures of association for use with ordinal dependent variables. *Social Forces 58*, 4, 1178–1194.

DeValve, M. (2015). *A different justice: Love and the future of criminal justice practice in America.* Durham, NC: Carolina Academic Press.

DeValve, M., Garland, T., & Wright, E. (2018). *A unified theory of justice and crime: Justice that love gives.* Lanham, MD: Lexington.

DeValve, M. (2020a). Defunding the ramparts and institutional theory: The master's tools will fell the master's house. *Journal of Community Safety and Well-Being 5*, 4, 138–143.

DeValve, M. (2020b). Reconsidering the response to mass violence: Meaning, choice, and human truth. *Journal of Theoretical & Philosophical Criminology 12*, 80–97.

DiMaggio, P., & Powell, W. (1983). The iron cage revisited: Institutional isomorphism and collective rationality in organizational fields. *American Sociological Review 48*, 2, 147–160.

Doane, D., & Seward, L. (2011). Measuring skewness: A forgotten statistic? *Journal of Statistics Education 19*, 2. www.amstat.org/publications/jse/v19n2/doane.pdf.

Fisher, R. & Yates, F. (1963). *Statistical tables for biological, agricultural, and medical research, 6d.* Edinburgh: Oliver & Boyd.

Glaser, B., & Strauss, A. (1967/2012). *The discovery of grounded theory: Strategies for qualitative research*. New Brunswick, NJ: Aldine/Transaction.
Goodman, L. & Kruskal, W. (1954). Measures of association for cross classification. *Journal of the American Statistical Association 49*, 732–769.
Gould, S. (1981/1996). *The mismeasure of man*. New York: W. W. Norton.
Guttman, L. (1941). The quantification of a class of attributes: A theory and method for scale construction. In Horst, P. (Ed.). *The prediction of personal adjustment*. New York: Social Science Research Council, pp. 321–348.
Hallberg, L. (2009). The "core category" of grounded theory: Making constant comparisons. *International Journal of Qualitative Studies on Health and Well-being 1*, 3. https://doi.org/10.1080/17482620600858399.
Kachigan, S. (1986). *Statistical analysis: An interdisciplinary introduction to univariate and multivariate methods*. New York: Radius.
Kviz, F. (1981). Interpreting proportional reduction in error measures as percentage of variation explained. *The Sociological Quarterly 22*, 413–420.
Leyes, C., Ley, C., Klein, O., Bernard, P., & Licata, L. (2013). Detecting outliers: Do not use standard deviation around the mean, use absolute deviation around the median. *Journal of Experimental Social Psychology 49*, 4, 764–766.
Lindley, D. & Scott, W. (1984). *New Cambridge statistical tables, 2d*. Cambridge University Press.
Meyer, J. & Rowan, B. (1977). Institutionalized organizations: Formal structure as myth and ceremony. *American Journal of Sociology 83*, 2, 340–363.
Quinney, R. (2000). *Bearing witness to crime and social justice*. Albany: SUNY Press.
Storch, J., & Panzarella, R. (1996). Police stress: State-trait anxiety in relation to occupational and personal stressors. *Journal of Criminal Justice 24*, 2, 99–107.
Székely, G., Rizzo, M., & Bakirov, N. (2007). Measuring and testing dependence by correlation of distances. *The Annals of Statistics 35*, 6, 2769–2794.
Wang, Y., Li, Y., Cao, H., Xiong, M., Shugart, Y, & Jin, Y. (2015). Efficient test for nonlinear dependence of two continuous variables. *BMC Bioinformatics 16*, 260–268.
Warner, R. (2008). *Applied statistics: From bivariate through multivariate techniques*. Thousand Oaks, CA: Sage.
Weisburd, D., & Britt, C. (2007). *Statistics in criminal justice, 3d*. New York: Springer.
Weisburd, D., & Piquero, A. (2008). How well do criminologists explain crime? Statistical modeling in published studies. *Crime and Justice 37*, 1, 453–502. https://doi.org/10.1086/524284.
Wilson, T. (1971). Critique of ordinal variables. *Social Forces 49*, 3, 432–444.
Young, J. (2011). *The criminological imagination*. Malden, MA: Polity.
Ziliak, S., & McCloskey (2011). *The cult of statistical significance: How the standard error costs us jobs, justice and lives*. Ann Arbor: University of Michigan Press.

Appendix A

Proportions of the Area under the Curve, Normal (z) Distribution (measured from the mean, 0)

Z	*0*	*0.01*	*0.02*	*0.03*	*0.04*	*0.05*	*0.06*	*0.07*	*0.08*	*0.09*
0	0.0000	0.0040	0.0080	0.0120	0.0160	0.0199	0.0239	0.0279	0.0319	0.0359
0.1	0.0398	0.0438	0.0478	0.0517	0.0557	0.0596	0.0636	0.0675	0.0714	0.0713
0.2	0.0793	0.0832	0.0871	0.0910	0.0948	0.0987	0.1026	0.1064	0.1103	0.1141
0.3	0.1179	0.1271	0.1255	0.1293	0.1331	0.1368	0.1406	0.1443	0.1480	0.1517
0.4	0.1554	0.1591	0.1628	0.1664	0.1700	0.1736	0.1772	0.1808	0.1844	0.1879
0.5	0.1915	0.1950	0.1985	0.2019	0.2054	0.2088	0.2123	0.2157	0.2190	0.2224
0.6	0.2257	0.2291	0.2324	0.2357	0.2389	0.2422	0.2454	0.2486	0.2517	0.2549
0.7	0.2580	0.2611	0.2642	0.2673	0.2704	0.2734	0.2764	0.2794	0.2823	0.2852
0.8	0.2881	0.2910	0.2939	0.2967	0.2995	0.3023	0.3051	0.3078	0.3106	0.3133
0.9	0.3159	0.3186	0.3212	0.3238	0.3264	0.3289	0.3315	0.3340	0.3365	0.3389
1	0.3413	0.3438	0.3461	0.3485	0.3508	0.3531	0.3554	0.3577	0.3599	0.3621
1.1	0.3643	0.3665	0.3686	0.3708	0.3729	0.3749	0.3770	0.3790	0.3810	0.3830
1.2	0.3849	0.3869	0.3888	0.3907	0.3925	0.3944	0.3962	0.3980	0.3997	0.4015
1.3	0.4032	0.4049	0.4046	0.4082	0.4099	0.4115	0.4131	0.4147	0.4162	0.4177
1.4	0.4192	0.4207	0.4222	0.4236	0.4251	0.4265	0.4279	0.4292	0.4306	0.4319
1.5	0.4332	0.4345	0.4357	0.4370	0.4382	0.4394	0.4406	0.4418	0.4429	0.4441
1.6	0.4452	0.4463	0.4474	0.4484	0.4495	0.4505	0.4515	0.4525	0.4535	0.4545
1.7	0.4554	0.4564	0.4573	0.4582	0.4591	0.4599	0.4608	0.4616	0.4625	0.4633
1.8	0.4641	0.4649	0.4656	0.4664	0.4671	0.4678	0.4686	0.4693	0.4699	0.4706
1.9	0.4713	0.4719	0.4726	0.4732	0.4738	0.4744	0.4750	0.4756	0.4761	0.4767
2	0.4772	0.4778	0.4783	0.4788	0.4793	0.4798	0.4803	0.4808	0.4812	0.4817
2.1	0.4821	0.4826	0.4830	0.4834	0.4838	0.4842	0.4846	0.4850	0.4854	0.4857
2.2	0.4861	0.4864	0.4868	0.4871	0.4875	0.4878	0.4881	0.4884	0.4887	0.4890
2.3	0.4893	0.4896	0.4898	0.4901	0.4904	0.4906	0.4909	0.4911	0.4913	0.4916
2.4	0.4918	0.4920	0.4922	0.4925	0.4927	0.4929	0.4913	0.4932	0.4934	0.4936
2.5	0.4938	0.4940	0.4941	0.4943	0.4945	0.4946	0.4948	0.4949	0.4951	0.4952
2.6	0.4953	0.4955	0.4956	0.4957	0.4959	0.4960	0.4961	0.4962	0.4963	0.4964
2.7	0.4965	0.4966	0.4967	0.4968	0.4969	0.4970	0.4971	0.4972	0.4973	0.4974
2.8	0.4974	0.4975	0.4976	0.4977	0.4977	0.4978	0.4979	0.4979	0.4980	0.4981
2.9	0.4981	0.4982	0.4982	0.4983	0.4984	0.4984	0.4985	0.4985	0.4986	0.4986
3	0.4987	0.4987	0.4987	0.4988	0.4988	0.4989	0.4989	0.4989	0.4990	0.4990

Adapted from Warner, R. (2007). *Applied statistics: from bivariate through multivariate techniques*. Thousand Oaks, CA: Sage.

Appendix B

Critical Values of t and Degrees of Freedom

df	*One-Tailed*					
	0.1	*0.05*	*0.025*	*0.01*	*0.005*	*0.0005*
	Two-Tailed					
	0.2	*0.1*	*0.05*	*0.02*	*0.01*	*0.001*
1	3.078	6.314	12.706	31.821	63.657	636.619
2	1.886	2.920	4.303	6.965	9.925	31.598
3	1.638	2.353	3.182	4.541	5.841	12.941
4	1.533	2.132	2.776	3.747	4.604	8.610
5	1.476	2.015	2.571	3.365	4.032	6.859
6	1.440	1.943	2.447	3.143	3.707	5.959
7	1.415	1.895	2.365	2.998	3.499	5.405
8	1.397	1.860	2.306	2.896	3.355	5.041
9	1.383	1.833	2.262	2.821	3.250	4.781
10	1.372	1.812	2.228	2.764	3.169	4.587
11	1.363	1.796	2.201	2.718	3.106	4.437
12	1.357	1.782	2.179	2.681	3.055	4.318
13	1.350	1.771	2.160	2.650	3.012	4.221
14	1.345	1.761	2.145	2.624	2.977	4.140
15	1.341	1.753	2.131	2.602	2.947	4.073
16	1.337	1.746	2.120	2.583	2.921	4.015
17	1.333	1.740	2.110	2.567	2.898	3.965
18	1.330	1.734	2.101	2.552	2.878	3.922
19	1.328	1.729	2.093	2.539	2.861	3.883
20	1.325	1.725	2.086	2.528	2.845	3.850
21	1.323	1.721	2.080	2.518	2.831	3.819
22	1.321	1.717	2.074	2.508	2.819	3.792
23	1.319	1.714	2.069	2.500	2.807	3.767
24	1.318	1.711	2.064	2.492	2.797	3.745
25	1.316	1.708	2.060	2.485	2.787	3.725
26	1.315	1.706	2.056	2.479	2.779	3.707
27	1.314	1.703	2.052	2.473	2.771	3.690
28	1.313	1.701	2.048	2.467	2.763	3.674
29	1.311	1.699	2.045	2.462	2.756	3.659
30	1.310	1.697	2.042	2.457	2.750	3.646
40	1.303	1.684	2.021	2.423	2.704	3.551
60	1.296	1.671	2.000	2.390	2.660	3.460
120	1.289	1.658	1.980	2.358	2.617	3.373
∞	1.282	1.645	1.960	2.326	2.576	3.291

Adapted from Fisher, R. & Yates, F. (1963). *Statistical tables for biological, agricultural, and medical research, 6d*. Edinburgh: Oliver & Boyd.

Appendix C

Area under the Curve, F Distribution, p = 0.05 values

DF for the Denominator	*DF for the Numerator*									
	1	*2*	*5*	*8*	*10*	*30*	*40*	*60*	*120*	∞
1	161.40	199.50	230.20	238.90	241.90	250.10	251.10	252.20	253.30	254.30
2	18.51	19.00	19.30	19.37	19.40	19.46	19.47	19.48	19.49	19.50
3	10.13	9.55	9.01	8.85	8.79	8.62	8.59	8.57	8.55	8.53
4	7.71	6.94	6.26	6.04	5.96	5.75	5.72	5.69	5.66	5.63
5	6.61	5.79	5.05	4.95	4.88	4.50	4.46	4.43	4.40	4.36
6	5.99	5.14	4.39	4.15	4.06	3.81	3.77	3.74	3.70	3.67
7	5.59	4.74	3.97	3.73	3.64	3.38	3.34	3.30	3.27	3.23
8	5.32	4.46	3.69	3.44	3.35	3.08	3.04	3.01	2.97	2.93
9	5.12	4.26	3.48	3.23	3.14	2.86	2.83	2.79	2.75	2.71
10	4.96	4.10	3.33	3.07	2.98	2.70	2.66	2.62	2.58	2.54
11	4.84	3.98	3.20	2.95	2.85	2.57	2.53	2.49	2.45	2.40
12	4.75	3.89	3.11	2.85	2.75	2.47	2.43	2.38	2.34	2.30
13	4.67	3.81	3.03	2.77	2.67	2.38	2.34	2.30	2.25	2.21
14	4.60	3.74	2.96	2.70	2.60	2.31	2.27	2.22	2.18	2.13
15	4.54	3.68	2.90	2.64	2.54	2.25	2.20	2.16	2.11	2.07
16	4.49	3.63	2.85	2.59	2.49	2.19	2.13	2.11	2.06	2.01
17	4.45	3.59	2.81	2.55	2.45	2.15	2.10	2.06	2.01	1.96
18	4.41	3.55	2.77	2.51	2.41	2.11	2.06	2.02	1.97	1.92
19	4.38	3.52	2.74	2.48	2.38	2.07	2.03	1.98	1.93	1.88
20	4.35	3.49	2.71	2.45	2.35	2.04	1.99	1.95	1.90	1.84
30	4.17	3.32	2.53	2.27	2.16	1.84	1.79	1.74	1.68	1.62
40	4.08	3.23	2.45	2.18	2.08	1.74	1.69	1.64	1.58	1.51
60	4.00	3.15	2.37	2.10	1.99	1.65	1.59	1.53	1.47	1.39
120	3.92	3.07	2.29	2.02	1.91	1.55	1.50	1.43	1.35	1.25
∞	3.84	3.00	2.21	1.94	1.83	1.46	1.39	1.32	1.22	1.00

Adapted from Kachigan, S. (1986), p. 566. Original source: Pearson & Hartley, eds. *Biometrika Tables for Statisticians*.

Appendix D

Chi-Square Distribution Values

df	*Probability*			
	0.1	*0.05*	*0.01*	*0.001*
1	2.706	3.841	6.636	10.827
2	4.605	5.991	9.210	13.815
3	6.251	7.815	11.345	16.266
4	7.779	9.488	13.277	18.467
5	9.236	11.070	15.086	20.515
6	10.645	12.592	16.812	22.457
7	12.017	14.067	18.475	24.322
8	13.362	15.507	20.090	26.125
9	14.684	16.919	21.666	27.877
10	15.987	18.307	23.209	29.588
11	17.275	19.675	24.725	31.264
12	18.549	21.026	26.217	32.909
13	19.812	22.362	27.688	34.528
14	21.064	23.685	29.141	36.123
15	22.307	24.996	30.578	37.697
16	23.542	26.296	32.000	39.252
17	24.769	27.587	33.409	40.790
18	25.989	28.869	34.805	42.312
19	27.204	30.144	36.191	43.820
20	28.402	31.410	37.566	45.315
21	29.615	32.671	38.932	46.797
22	30.813	33.924	40.289	48.268
23	32.007	35.172	41.638	49.728
24	33.196	36.415	42.980	51.179
26	35.563	38.885	45.642	54.052
28	37.916	41.337	48.278	56.893
30	40.256	43.773	50.892	59.703
32	42.585	46.194	53.486	62.487
34	44.903	48.602	56.061	65.247
36	47.212	50.999	58.619	67.985
38	49.513	53.384	61.162	70.703
40	51.805	55.759	63.691	73.402
44	56.369	60.481	68.710	78.750
50	63.167	67.505	76.154	86.661
54	67.673	72.153	81.069	91.872
60	74.397	79.082	88.379	99.607
64	78.860	83.675	93.217	104.716
70	85.527	90.531	100.425	112.317

Adapted from Fisher, R. & Yates, F. (1963). *Statistical tables for biological, agricultural, and medical research, 6d*. Edinburgh: Oliver & Boyd.

Index

Note: **Bold** page numbers refer to tables and *italic* page numbers refer to figures.

addition rule 56–57, *59*
agency data 4, 73, 109
agency statistician 10
alpha inflation, ANOVA 116–118
alternative hypothesis 91, 96, 98
analysis decision process *2*, 2–8
ANOVA (analysis of variance): alpha inflation 116–118; basic logic of 118–119; Bonferroni correction 125; calculation 124; degrees of freedom between 120; degrees of freedom within 120–123; one-way ANOVA 119; post-hoc analyses 124–125; sum of squares, between-group 119–120; sum of squares, partitioning of 123–124; within-group sum of squares 120
areas under the curve (AUC) 64
Argyris, C. 3
arithmetic mean 29, 31, 32, 33, 37, 41
astrophysics analogy *77*
asymptotic probability *79*
AUC *see* areas under the curve (AUC)

bar graphs 20–22
basic data presentation techniques: cumulative frequency 13–14; cumulative percent 17–18; percents 16–17; ratios and proportions 14–16
Bernoulli experiment 59, 67
binomial distribution 57–65, 78
binomial probability formula 67
binomial theorem 175
biological adaptation 88
bivariate hypothesis testing: independent-samples *t* test 107–111; paired-samples *t* test 112–115; single-sample *t* test 103–107
bivariate linear regression 146–151; example of 151–155
Bonferroni correction 125
Britt, C. 136, 140
broad-class/top-down theory 88

CANOVA 142
causation 89–90
CED *see* classical experimental design (CED)
centrality: mean 29–31; median 28–29; mode 27–28; pythagorean means 32–37; weighted mean 31–32
Central Limit Theorem 72, 80, 86, 94, 116, 146, 158; descriptive and inferential statistics 66–67; mean, sampling distribution of 69–75; reasoning toward inference 67–69
chi square (χ^2) 159–160; calculation of 160–163; interpretation 163; phi and Cramer's V 163–165
City of Detroit's Open Data Portal 10
classical experimental design (CED) 67
coefficient of variation (CV) 48–49
Cohen, J. 99
comparison process 13, 14, 61
complement, probability 56
concerning hypotheses 87–89
conditional probability 56–57; addition rule 56–57; capacity 57; multiplication rule 57; replacement and non-replacement 56
confidence intervals 80–82, 83–84
context 13, 61, 66
continuous variables 22, 23, 27
correlation 89–90; logic of 128–129; Pearson *r* 130–142; phi (Ø) 143–144; point biserial, r_{pb} 143; Spearman ρ 142–143
correlation matrix 128, **129**
Costner, Herbert 157
covariance 135–137, **137**,
creating process 10
crime rates 17
criminal event 11
criminological statistics 10

criminological theory 3, 87
criminology 94
criteria 8
cumulative frequency 13–14
cumulative percent 17–18

databases 5
data, graphic presentation: bar graphs 20–22; continuous variables 23; histograms 22–23; pie charts 18–20; polygons 23–24; theoretical continuity 23
decision-making 4
decision-support frameworks 8
degrees of freedom 43, 55, 83, 108
descriptive statistics 66–67
different skew environments *50*
directional hypothesis 92
directionality 92
direct relationship 89
distributional normality 132–135
distributions 24–25, 44, 49
Dream Theater 8

education 5
educational attainment 5
effect size 96–99
empirical distributions 24
epistemology 88
EPSEM (Equal Probability of SElection Method) 55
equal probability assumption 68
equal units 29
error grid 95–96, *96*
errors types 95–96; type I error 95–96; type II error 96
eta squared 111–112
etiological why 3
etiology 3, 88
event 55
experiment 55
explained variance 155

factorial ANOVA 117
falsification 90–91
Fayetteville (NC) Police Department (FPD) 1
Fisher–Pearson skewness coefficient 50
Frankl, Viktor 61
frequency table 11

gamma (γ) 168–174
Gaussian curve 61, 65
geometric mean 32–33, 34
Goodman, L. 157, 166–174
Gossett, William Sealy 102
GPS device's software 88
grounded theory 86

Hallberg, L. 8
hamartioetiology 88
HARKing 86
harmonic mean 35–36
heteroscedasticity 132
histograms 22–23, 44
homework assignments 31
homoscedasticity 132
human scientific research project 2
hypothesis 3, 87, *87*; alternative hypothesis 91; basic anatomy of 89; directional hypothesis 92; first axis 87–88; nondirectional hypothesis 92; null hypothesis 91; one-tailed hypothesis 95; second axis 88–89; sets 92; testing 86; two-tailed hypothesis 95

imputing data 12
independence 130
independent-samples *t* test 107–111
independent variable 102
inferential statistical tests 66–67, 77, 86
insurance policies 17
"interval" level, measure 7
inverse relationship 89

Kachigan, S. 69
Kruskal, W. 157, 166–174
kurtosis 51–52
kurtotic data 52

lambda *(λ)* 166–167
leptokurtic distributions 51
Levene's test 110
Leyes, C. 48
linear relationship 142

MAD *see* mean absolute deviation (MAD)
Mann–Whitney *U* 143
margin of error 81
McCloskey, S. 93, 100
mean 29–31
mean absolute deviation (MAD) 40–43
mean deviation 41
meaningfulness 61, 63, 65, 66, 93
mean square 119
measure decision tree *8*
measure levels 6
median 28–29, 47–48
median sign test 174–176
Metz, Johann 61
miles-per-hour metric 37
mode 27–28
monotonic relationship 142
multiplication rule 57
mutually exclusive outcomes 56–57, 92
μ(Mu), population parameter 76, 77

narrow-class/bottom-up theory 88
National Policing Institute (NPI) 2
nondirectional hypothesis 92
nonlinearity *131*
non-numeric information 5
nonparametric tests 158
non-replacement probability 56
nonspuriousness 90
normal curve 57–65
normal distribution 49, 62
normal forms 5
normalization 5
NPI *see* National Policing Institute (NPI)
null hypothesis 91, 94, 110, 142, 154, 161

observed outcome 63
one-tailed fashion 106
one-tailed hypothesis 95
one-way ANOVA 117, 119
ontology 88
Operation ID efforts 19
ordinal data 28
Ordinary Least Squares 57
outliers 47–48

paired-samples *t* test 112–115
paired scores 132
PDI *see* Police Data Initiative (PDI)
Pearson correlation 155
Pearson, Karl 49
Pearson *r:* calculation 135–140; distributional normality 132–135; independence 130; linearity 130–132, *131*; linear relationship 142; measurement, level of 132; nonlinearity *131*; paired scores 132; significance of 140–142
percents/percentage 16–17
phi (Ø) 143–144
pie charts 18–20
point biserial correlation 111–112
point biserial, r_{pb} 143
point estimates 77–78; confidence intervals 80–82, 83–84; degrees of freedom 83; probability distributions and inference 78–80; student's *t* varies 83–84; *t* distribution 82–83; *z* distribution 82–83
Police Data Initiative (PDI) 1–2, 73
polygons 23–24
populations 54; identity 68; mean 104
position, modal value 28
post-hoc analyses 124–125
prediction 147, 153
presentation options 25
presenting data: basic data presentation techniques 10–18; data, graphic presentation of 18–24; distributions 24–25; presentation options 25; uncertainty 24–25; variability 24–25
President's Task Force 2
primary data 4
priori theory 129
probability 55; binomial distribution 57–65; complement of 56; conditional probability 56–57; normal curve 57–65; observed/empirical outcome 60; outcomes, expected set of 60; probability curve *60*
probability density functions 62, 66, 67
probability density histogram 70
probability distributions 78–80
probability theory 25, 54
program theory 3
property crimes 12, **12, 13, 16,** 17, *18,* 19, 54
proportionate reduction in error (PRE) 165–166; Goodman and Kruskal's Gamma 168–174; Goodman and Kruskal's Lambda 166–167; measures for nominal variables 166–174; median sign test 174–176
proportions 14
"protected" analyses *see* post-hoc analyses
Pythagorean mean family: geometric mean 33; harmonic mean 35–36; pythagorean means 32
Pythagorean means 32–37

qualitative data 4–5
quantitative data 4

randomness 54
range 39–40
"Ratio of Burglaries to Larcenies" 14
ratios 7, 14–16
reasoning behind hypothesis testing: causation 89–90; concerning hypotheses 87–89; directionality 92; error grid 95–96; errors types 95–96; falsification 90–91; hypothesis sets 92; scientific reasoning 90–91; statistical decisions 90–91; statistical significance 93–95
rejection regions *95*
relative frequency 23, 55
religious preference 27, 40, **41**
repeated-measures ANOVA 117
replacement probability 56
research question 3

samples 54
sample space 55
sampling distribution 69, 76, 77, 80, 104; of mean 69–75
Schon, D. 3
scientific reasoning 90–91
secondary data 4
signal-to-noise dynamic 51

signal-to-noise ratios 6, 45, 119, 146
single-sample *t* test 103–107
skew/skewness 49–51; Fisher–Pearson coefficient 50; Pearson's second coefficient of 50; and transformations **52**
SMC *see* stress management class (SMC)
software programs 6
Spearman ρ 142–143
spreadsheet 5, 66
squared deviation **46**
stakeholders 4
standard deviation 66
standard errors 103, 107; estimation 72, **74**, 80, 84
statistical analytical techniques 10
statistical decisions 90–91, 110, 115
statistical power 96–100, *97*
statistical reasoning 9
statistical significance 93–95, *95*
statistical techniques 93
statistical tests 102
stress management class (SMC) 121
student's *t* distribution 102
student's *t* varies 83–84
sum of squared errors 150
sum of squares 119, **123**; between-group 119–120; partitioning of 123–124

t distribution 82–83
temporal order 90
theoretical continuity 23
theoretical distribution 24
theory-application axis 87
"theory of action" 3
Tillich, Paul 61
transformations 52, **52**
true dichotomous variable 143
t test 103
Twain, Mark 10
two-tailed hypothesis 95
typicality 27, 35

uncertainty 24–25
unit of analysis 5

variability 24–25; coefficient 48–49; kurtosis 51–52; kurtotic data 52; mean absolute deviation 40–42; outliers 47–48; range 39–40; skew 49–51; standard deviation 42–43; variation ratio 40; z score 43–47, **44**
variance, homogeneity of 110–111
variation ratio 40
volatility 47

Wang, Y. 142
Warner, R. 50, 143
weighted mean 31–32
Weil, Simone 61
Weisburd, D. 136, 140
Wilcoxon Rank Sum test 143
Wilson, T. 157
wisdom-abandoned behavior 89
Wizard of Oz 3
Wright, Elizabeth Dr. 1

Y intercept 147

z distribution 82–83
zemioetiology 88
Ziliak, S. 93, 100, 157
z scores 43–47, **44**, **47**, 63

Milton Keynes UK
Ingram Content Group UK Ltd.
UKHW052227100624
443929UK00007B/68

9 781032 638546